Spreadsheet Exercises and Applications for Business Math

Toby Wraye, Ed. M.

With technical review by Lyndon Hebenstreit

Prentice Hall, Upper Saddle River, NJ 07458

Production Supervisor: *Mary Carnis*
Director of Publishing: *Bruce Johnosn*
Production Manager: *Ilene Sanford*
Production Editor: *Barbara A. Cassel*

A Simon & Schuster Company
Upper Saddle River, NJ 07458

Printed in the United States of America

10 9 8 7 6 5 4 3 2 1

ISBN 0-13-486739-4

Prentice-Hall International (UK) Limited, *London*
Prentice-Hall of Australia Pty. Limited, *Sydney*
Prentice-Hall Canada Inc., *Toronto*
Prentice Hall Hispanoamericana, S.A., *Mexico*
Prentice-Hall of India Private Limited, *New Delhi*
Prentice-Hall of Japan, Inc., *Tokyo*
Simon & Schuster Asia Pte. Ltd., *Singapore*
Editoria Prentice-Hall do Brasil, *Ltda., Rio de Janero*

CONTENTS

V. OTHER EXPENSES

Preface

This workbook is designed primarily to introduce students to the use of spreadsheets as they might be used in real life, business situations. As students work through the text, exercises, and applications they will have hands-on experience with spreadsheet design principles, data entry techniques, business math formulas, and be introduced to the use of common spreadsheet functions.

For each section of the workbook students are asked to follow step-by-step directions and create a template they can later apply to daily use. Of course, as they move from the academic world they will find each business handles things somewhat differently than this text could expect. With this experience, however, they should be prepared to adapt to similar situations.

This text is based on a self-paced learning environment that lets students gauge when they are ready to move on from chapter to chapter. They are advised to move from chapter 1 to 18 in that order, but it is entirely appropriate for them to skip around as their interests lead them. A section-by-section process is probably more ideal.

This workbook also benefits from the support of existing classroom text. Most of its principles and examples were taken from three other popular and respected Prentice Hall business math texts: *Business Math* by Cheryl Cleaves and Margie Hobbs; *Business Mathematics* by Nelda W. Roueche and Virginia H. Graves; and *Business Mathematics* by Esther H. Highland, Roberta S. Rosenbaum, and Charles Peselnick. As students tackle the exercises and applications, they may well recognize the examples from their regular classroom text.

Throughout this workbook the author has employed an approachable readable style that puts the emphasis on simple and direct reporting. Some effort has gone into explaining the more complex aspects of business math but the text remains fundamentally a spreadsheet instruction book rather than a business math book. It is expected the student will benefit from a more formal business math class and use this text in the computer lab.

The author has relied mainly on Microsoft Excel for Windows and the text's instructions often reflect that bias. However, it is expected that students using Lotus 1-2-3 for Windows or Microsoft Excel for the Macintosh will be fully comfortable with the directions. Students should have had an introduction to spreadsheets in class before tackling this text. The first section covering three chapters is intended to introduce them to spreadsheet use as well as to business math terminology and theory.

The author wishes to thank Lyndon Hebenstreit for his exceptional consultation and contribution to this endeavor. Without his support, encouragement, and expert input this text would not have been possible. Mr. Hebenstreit is a business consultant and accountant, living, teaching, and serving his clients in Ojai, California, as owner of the Quick Booking bookkeeping and business consultation firm.

The author also wishes to acknowledge the support and care his partner and children have provided during the long hours it took to prepare this text. He also wishes to dedicate this work to his mother, Jane Ann Burdick who at 74 continues to embrace life with dignity and humor.

Toby Wraye

I. Introductory Skills

1. Using a Spreadsheet: Review

> This workbook is intended for students already familiar with basic spreadsheet usage, however this first section reviews the skills needed to complete the exercises in the following sections and introduces some common math using spreadsheets. Much of the spreadsheet functionality is not covered in this chapter. The following chapters in this section introduce more about spreadsheet use and design as well as business math concepts.

Objectives

By working through and completing the exercises below, you will be able to use a spreadsheet to:

- Perform data entry and basic math.
- Format data and text.
- Use references and prepare formulas.
- Use functions in formulas.
- Create graphs and charts.

Data Entry

Spreadsheets present the user with rows and columns that intersect to form cells. Three types of data can be entered into a spreadsheet cell: text data (labels), numeric data (value), and formulas. Here you will review text and numeric data entry.

Starting and Saving a Spreadsheet

Like most programs, when you open or execute a spreadsheet program it presents a new file. This file exists in RAM only. You can create a new spreadsheet file from this opening file by selecting the Save or Save As command and naming the file. You can view existing spreadsheet files by selecting the Open or Retrieve command. You can also have the program create a new spreadsheet file by giving the New command.

Different spreadsheet programs have different names for commands but the functionality remains fairly consistent from one program to another. Once you learn one you can probably find your way around another. You can review the commands for Microsoft Excel and Lotus 1-2-3 for Windows in the Appendices of this workbook. (This workbook and its examples are based primarily on Microsoft Excel for Windows version 5.0; however, it is generic enough to allow you to use your program of choice.)

Entering Labels

Text data is called a label because, in a spreadsheet, text is most often used to label a row or column's values. Text can either be a letter of the alphabet or a number used as text (nominal numbers). Usually a spreadsheet program enters text aligned left meaning the text begins at the left edge of a cell. All letters and most symbols are treated as text. You can format a cell so its text aligns right or centered and even centered between the top and bottom of a cell. When you use labels as headings, you can format the cell so that the text is centered across a range of cells. You can also have the text in a cell "wrap" at the cell's edges as if the edges were the margins of a page.

Exercise 1

To enter labels: move the focus to (ENTER, TAB, ARROW KEYS) or select (left-mouse-click) a cell, enter text from the keyboard, and press ENTER. Prepare the labels for your first spreadsheet.

1. Open your spreadsheet program.
2. Save the opening spreadsheet as the PRACTICE file.
3. Cell A1 should be selected: that is, it has the focus and what you enter will show up in this cell. Enter the title of this spreadsheet: **First Practice Spreadsheet**.
4. Starting in cell B3 and moving across the row to cell E3, enter the column labels of: **Student Project 1, Student Project 2, Student Project 3, Average Score**.
5. Select cells A1 to E1 and choose the Center Across Selection command (Format, Cell, Alignment, for example). The title appears centered above the labels.
6. Save the spreadsheet.

The labels in cells B3 to E3 are truncated, so the cells either need to be widened or the text wrapped within the boundaries of the cell. They are to be column headings as well, so they should show up darker than other text. To increase the width of the cells or wrap the text and bold the cell's contents:

1. Select cells B3, C3, D3, and E3.
2. Find and choose the AutoFit command (Format, Column, Width, AutoFit, for example). The columns widen to the width of the entries.
3. Find and choose the Standard Width command. The columns return to the default width.
4. Find and choose the Wrap command (Format, Cell, Alignment, Wrap Text, for example). The text wraps within the cells.
5. Choose the Bold command (Format, Cell, Font, Bold).
6. Save the spreadsheet (File, Save).

Creating a Series and Removing Cell Entries

Spreadsheets have the capability to enter text and numbers for you. If the text down a column or across a row is a series with a clear pattern to it, all you need to do is enter the first two or three entries, select the range of cells that will hold the data, and choose the Series command.

Removing data is a simple matter of selecting a cell or a range of cells and pressing the DELETE key. You can also use hot keys or menu commands to remove entries. Entering new text also removes existing entries.

Exercise 2

Complete the structure of the PRACTICE spreadsheet by entering a series of labels and then remove some labels:

1. Enter the following text in cells A4 and A5: **Year 1, Year 2**.
2. Select the range from A4 to A8.
3. Find and choose the Series command (Edit, Fill, Series, Autofill, Increment by 1). Year 3, Year 4, and Year 5 appear in the remaining cells.
4. Format the column for AutoFit
5. Bold the text.
6. Select cell A7 and press the DELETE key or choose the Delete command (Edit, Clear).
7. Select cell A8 and enter Total. The program replaces the text and keeps its formatting.
8. Save the spreadsheet.

Entering and Formatting Values

In spreadsheets, numeric data are called values. Most spreadsheet programs place values right aligned in their cells. You can format values in any way appropriate to their meaning. For example, you can enter 1.5 as currency ($1.50) or as a percent (150%) or as a fraction (1 1/2). You can enter 6/4 as a date (June 4, 1996). It is very common in business math to see currency, dates, percents, and decimals. Sometimes fractions are used.

Exercise 3

To enter and format a value:

1. Open the PRACTICE file
2. In cell B3 enter 1514.67.
3. Press the ENTER key. The number 1514.67 appears in cell B3 and the focus moves to the next cell (depending on how your spreadsheet program's preferences are set, ENTER moves down one cell or right one cell).
4. Select cell B3 and choose the format command for formatting currency.
5. Select the ($#,##0.00_);($#,##0.00) number style to have this value show up as currency with commas and two (hundreds) decimal places.
6. Exit the formatting dialog box and the cell's number presents as currency with a dollar sign, commas, and decimal places in the hundredths place. (If all you see is ##########, then widen the cell to accommodate the width of the figure.)
7. Select cell A4, which now has Year 1.
8. Enter 6/1 and press Enter. A date should show up in the cell.
9. Select cell A4 and choose the format command for formatting date and time.
10. Choose the mmm-yy style. The date changes to Jun-96 (the year will be determined by the computer).
11. Save the spreadsheet.

Exercise 4

Add the following data (Table 1-1) to the PRACTICE spreadsheet and then format each number as it appears in the list. Notice that you only need to type over existing values and labels to replace them with new values or labels. Save the spreadsheet.

Table 1-1

Cell		Cell		Cell		Cell	
A3	**Jun-96**	B3	$1,514.67	C3	$2,289.75	D3	$3,589.77
A4	**Sep-96**	B4	$1,798.83	C4	$2,783.12	D4	$4,145.99
A5	**Dec-96**	B5	$1,895.34	C5	$3,347.23	D5	$3,973.45

Tip: *You can select a range from B3 to D5 or select columns A and then B to D and format all the similar values at one time.*

Your spreadsheet should look like Figure 1-1.

A1 | First Practice Spreadsheet

	A	B	C	D	E
1		**First Practice Spreadsheet**			
2		**Student Project 1**	**Student Project 2**	**Student Project 3**	**Average Amount**
3	**Jun-96**	$1,514.67	$2,289.75	$3,589.77	
4	**Sep-96**	$1,798.83	$2,783.12	$4,145.99	
5	**Dec-96**	$1,895.34	$3,347.23	$3,973.45	
6					
7	**Total**				
8					

Figure 1-1: Entering text and data in a spreadsheet.

Sorting Data

A final consideration in formatting data is the order in which it is presented. Spreadsheets are capable of rearranging data according to the order you specify. Labels can be sorted alphabetically in ascending or descending order and values are sorted in numeric ascending or descending order. You need to select the cells or whole rows or columns and then choose the Sort command (Data, Sort).

References and Formulas

A formula is made up of values and operators. The operators (see Table 1-2) tell the program what to do with the values and in what order to do it. Each value's operator precedes the value and the whole formula is preceded by an operator. You can "group" formulas within other formulas to control the order or allow the program's default order of operations to group them. For instance, =14-8*26/39 equals 8.67 but =(14-8)*26/39 equals 4 because the (14-8) is calculated first instead of last.

Table 1-2

Operator	Use
+	Addition (+3 means increase the value of this cell by 3).
-	Subtraction (-3 means decrease the value of this cell by 3).
*	Multiplication (*3 means increase the value of this cell 3 times).
/	Division (/3 means distribute the value of this cell into 3 portions).
^	Power or Exponent (+3^2 means to multiply 3 times itself two times; 3 X 3 X 3).
( )	Isolates a formula nested within a formula.

Entering Formula Data

Although you can easily enter =(14/7)*6 in a cell to see the answer 12, this is not the most efficient use of spreadsheets. Instead it is best to enter values into cells and then reference those cells in a formula. A reference refers to a cell's column letter and row number (for example, B3 is column B, row 3). In that way you need only enter a formula once, and you can change the cell's contents to see different results.

Other advantages become apparent when you copy formulas. Because a reference is relative to the location of the formula, when you copy formulas, the spreadsheet maintains this relationship. It does so by rewriting the moved formula to reference cells in the same relative position as were the original cells.

Exercise 5

To enter and copy a formula:

1. Open the PRACTICE spreadsheet.
2. In cell B8 enter =B4+B5+B6. The sum of the values in these cells appears in the Total row.
3. Select cell B8 and copy it (CTRL+C).
4. Select cell C8 and paste (CTRL+V) B8's data into it. The sum of the values in C4 to C6 appear in the total row. Select C8 again and you can see the formula in the command line.
5. Select cell B8 through D8 and choose the command to copy right (Edit, Fill, Right for example) B8's formula.
6. Save the spreadsheet.

Exercise 6

Enter the following formula to find the average values across the columns B, C, and D in row 4: =(B4+C4+D4)/3. Copy this formula down to row 6. Save the spreadsheet.

Your spreadsheet should look like Figure 1-2.

A1 | First Practice Spreadsheet

	A	B	C	D	E
1		First Practice Spreadsheet			
2		Student Project 1	Student Project 2	Student Project 3	Average Amount
3	Jun-96	$1,514.67	$2,289.75	$3,589.77	$2,464.73
4	Sep-96	$1,798.83	$2,783.12	$4,145.99	$2,909.31
5	Dec-96	$1,895.34	$3,347.23	$3,973.45	$3,072.01
6					
7	Total	$5,208.84	$8,420.10	$11,709.21	
8					

Figure 1-2: Spreadsheet formulas using references give instant results no matter how often the data changes.

Absolute References and Using the Insert Command

You can control how formulas reference cells when copied or moved. You can do so by making the cell references relative, or absolute to the column, or row, or both. Then, when you copy or move a formula, and it will always reference the appropriate cell or range of cells. By default, all cell references are relative.

Very often when designing a spreadsheet you need to insert a column, row, or cell. Whenever you do so the spreadsheet moves the other columns, rows, or cells and adjusts the relative references

Exercise 7

To make a reference absolute and to find the portion of the total each part of Project 1 represents.

1. Open the PRACTICE spreadsheet.
2. Select column C or a cell in column C and choose the Insert Column command. The C column data moves to the right and an empty column C appears.
3. In cell C3 enter the label **Portion of Total 1**. Format it as centered in the cell.
4. In cell C4 enter the formula =B4/$B$8 and copy this down to B5 and B6. Select cells B5 and then B6 and notice that the first reference changed as you copied but the second did not.
5. Format B4 to B6 to show just two decimal places (Format, Cell, Number).
6. Save the spreadsheet.

Exercise 8

Insert columns and make headings to find the portions for Projects 2 and 3. In cell C4, change the absolute reference $B$8 to B$8, so it is only absolute to the row. Copy this formula to the new columns. Save the spreadsheet, which should now look like Figure 1-3 except for the value yousee in cell H8. (The value in cell H8 will be explained below.)

A1 | First Practice Spreadsheet

	A	B	C	D	E	F	G	H
1		First Practice Spreadsheet						
2		Student Project 1	Portion of Total 1	Student Project 2	Portion of Total 2	Student Project 3	Portion of Total 3	Average Amount
3	Jun-96	$1,514.67	29%	$2,289.75	27%	$3,589.77	31%	$2,464.73
4	Sep-96	$1,798.83	35%	$2,783.12	33%	$4,145.99	35%	$2,909.31
5	Dec-96	$1,895.34	36%	$3,347.23	40%	$3,973.45	34%	$3,072.01
6								
7	Total	$5,208.84		$8,420.10		$11,709.21		$8,446.05
8								

Figure 1-3: References make it possible to copy formulas and control what column and row value will be used.

Fractions, Ratios, Decimals, Percents, and Division

Fractions, ratios, decimals, percents, and division are all ways of distributing things among or across something else. They have to do with portioning. They differ in their purpose and application.

Addition, subtraction, and multiplication are easily accomplished using a spreadsheet and are easily understood. Division is not difficult for the spreadsheet but it can be difficult to understand when talking about fractions, ratios, etc. You can eliminate the confusion if you remember the concept of a reciprocal and look at every number as a division problem or fraction of its own. So when you see the number five (5) think of it as five divided by one (5/1) and instead of division, think of it as multiplying using the reciprocal (1/5) or the reciprocal's decimal form (.2).

Ratios are the most confusing members of this portion family. They are most easily understood by the cook's dilemma: If the recipe calls for 2 eggs to serve 5 people, how many eggs do I use for 25 people? To find the answer you set up a proportion statement: 2 is to 5 then what is to 25? (2/5=x/25). To solve this equation for x, you must multiply both sides by 25. Multiplying 2/5 by 25 is the same as multiplying .4 times 25.

Exercise 9

To illustrate the relationship between fractions, ratios, decimals, percents, and division:

1. Open the PRACTICE spreadsheet.
2. In any empty cell enter =2/5. By default most spreadsheets will express the answer as a decimal.
3. Format the cell as a fraction and then as a percent.
4. Format the Portion column values as percents.
5. Save the spreadsheet. It should resemble Figure 1-3, except that now the Portions are percents.

Using Functions and Ranges

All spreadsheet programs have a variety of built-in functions. A function is like a small program. When you use it, it performs one of many formulas. Some of these are very elaborate algorithms, which are series of formulas applied to data and performed in a specific order to gain a single result. To enter a function you must use its name and then enter its arguments. The arguments identify the data upon which the function must act.

Functions often act on a range of data. A range is a special kind of reference that refers to more than one consecutive cell either down a column or across a row. For instance the range B3 to B7 includes cells B3, B4, B5, B6, and B7. The range C6 to G6 includes cells C6, D6, E6, F6, and G6; and the range B3 to D6 includes B3, B4, B5, B6, C3, C4, C5, C6, D3, D4, D5, and D6. You can refer to a range in a formula by using its first and ending cell references with a colon between them. Thus the range B3 to D6 is referred to as B3:D6. Most spreadsheets also allow you to name ranges (Insert, Name), which helps you to locate data and supports entering several ranges in a formula by simply typing the name rather than trying to remember their cell references.

Math and Statistic Functions

We will use just one math function throughout this book. The book also only makes use of two statistic functions. Addition is handled by the SUM function. You will use the AVERAGE function here and in the next chapter you will use the COUNT function.

Functions have two parts, the name of the function and its arguments. The two used here are:

Total =SUM(Range)

Average =AVERAGE(Range)

Exercise 10

Use the SUM and AVERAGE functions in place of the formulas in the PRACTICE spreadsheet.

1. Open the PRACTICE spreadsheet.
2. In cell B8 enter the SUM function and its arguments. In our example it is: =SUM(B4:B7). Notice that it includes the blank row. This technique allows you to add rows for new data and the range will automatically include the new rows.
3. Copy this formula to D8 and F8.
4. In cell H4 enter the formula to find the average amount for Jun-96: =AVERAGE(B4,D4,F4). Notice that we could not use a range because the cells are not contiguous.
5. Copy this formula down the column to cell F7. The formulas work for rows 5 and 6 but the seventh row reports a divide-by-zero error because there are no values to use in the referenced cells. (You will address this problem with a logical function below).

6. Your spreadsheet should look like Figure 1-3 except for the portions as percents the #DIV/0! (divide by zero) error message in cell H7.

7. Save the spreadsheet.

Logical Functions

Math functions help speed up complex calculations but sometimes you need to do one thing under one circumstance and another under different circumstances. In such a case you can use one of the logic functions. This workbook makes use of the IF, AND, and OR logical functions. They are used, for example, in the syntax:

=IF(X, then Y, otherwise Z) or *=IF(AND(A,B,...), then Y, otherwise Z)*

Lookup and Reference Functions

Lookup and reference functions bring text or a value from one location in the spreadsheet to be used in a formula or simply presented in a cell. For instance, the ISBLANK function tells you or a formula that the cell reference in the ISBLANK function's argument is or is not blank. It presents or "returns" TRUE if the cell is blank and FALSE if the cell has a label, value, or formula entered in it.

In this workbook you will also use the following lookup and reference functions: OFFSET to use the contents of a cell next to a range; INDEX to use the contents of a cell in an array (a group of cells); VLOOKUP to use the contents of a cell in a column; MATCH to return the position of the given value or text, that is, the column number in an array ("b" is 2 in the array "a", "b", "c"); COUNT to return the number of cells in a range.

Exercise 11

Eliminate the #DIV/0! Error in the Average Amount column and still keep the formula in the cell:

1. Open the PRACTICE spreadsheet.

2. Select cell H4.

3. Enter the following IF function which has the ISBLANK function nested within it: =IF(ISBLANK(Student Project 1), then "", otherwise AVERAGE(B4,D4,F4)), which in our example is, =IF(ISBLANK(B4),"",AVERAGE(B4,D4,F4)).

4. Copy this formula down to include cell H7. The formula returns a blank space rather than the divide by zero error.

5. Save the spreadsheet which still looks like Figure 1-3 except for the portions as percents and the blank space in cell H7.

Exercise 12

Extend the formulas for the Portion columns to row 7 but first edit them so the divide by zero error will not show.

Time and Date Functions

Time and date functions return a serial number of a date and time. The spreadsheet program recognizes time in terms of a number representing a day. There is a number for every day from the year 1900 to the year 2078. The TODAY() function returns the serial number and displays it as month, day, and year for the system's date (your computer has a calendar and clock). The DAY function returns the number of date so that =DAY(Jun 4, 1996) returns 4. The DAYS360 function converts a date to what it would be based on a 360 day year (12 months of 30 days each). The MONTH function returns the month of the referenced date.

Exercise 13

Add today's date to the PRACTICE spreadsheet.

1. Open the PRACTICE spreadsheet.
2. Select row 2 or a cell in row 2 and insert a row. The row is added above row 2 and the data is moved down.
3. In cell A2 enter the TODAY() function. It does not use arguments but you must include the parentheses. It appears formatted just as A1 was formatted: bold and centered across columns.
4. Save the spreadsheet. It should now look like Figure 1-4

A2 =TODAY()

	A	B	C	D	E	F	G	H
1				First Practice Spreadsheet				
2				7/22/96				
3								
4		Student Project 1	Portion of Total 1	Student Project 2	Portion of Total 2	Student Project 3	Portion of Total 3	Average Amount
5	Jun-96	$1,514.67	29%	$2,289.75	27%	$3,589.77	31%	$2,464.73
6	Sep-96	$1,798.83	35%	$2,783.12	33%	$4,145.99	35%	$2,909.31
7	Dec-96	$1,895.34	36%	$3,347.23	40%	$3,973.45	34%	$3,072.01
8								
9	Total	$5,208.84		$8,420.10		$11,709.21		$8,446.05
10								

Figure 1-4: The finished PRACTICE spreadsheet includes many of the features and uses many of the spreadsheet skills you will use in business math.

Accounting Functions

The final category of functions important to business math are the accounting functions. This workbook introduces you to three of them. The FV function returns the Future Value of an investment or loan given the interest rate, number of payment periods, and the payment amount. The PV function returns the Present Value of an investment or loan given the interest rate, the number of periods, and the payment amount. The PMT function returns the payment you would need to make on a regular basis to reach a future value given the interest rate, number of payment periods, and the present value.

Graphs and Charts

People will use the words *graph* and *chart* interchangeably but this text will refer to graphs as the drawing on a grid showing data. A chart is a display that includes the graph and explanatory text. The chart is a picture of the graph. The graph includes a horizontal- or x-axis (line running across) and a vertical- or y-axis (line running upwards). The x and y axes meet at 0.

Most spreadsheet programs include graphing and charting capabilities, or an "add-in" program is available to the program. To make a graph or chart you must first select a range and then choose the graph or chart command. The range you choose will depend on what you want to graph and what you want to show to the reader.

Bar Chart

A Bar chart shows adjacent lines that represent the different values in a range. The values are not necessarily related across time but could be. The three amounts for Jun 96 in the PRACTICE spreadsheet would compare Project 1, Project 2, and Project 3 figures with currency on the x-axis and Project 1, Project 2, and Project 3 on the y-axis.

Line Chart

A Line chart commonly shows progress over time. It usually has time on the x-axis. Time might also be represented by events such as the regular purchase of equipment or increases in employee salaries. The three values in the Average Amount column of the PRACTICE spreadsheet would logically be displayed in a line chart.

Pie Chart

A Pie chart displays a circular graph with "slices" showing what portion of a range each value represents. The three values under Project 1 in our example could be shown, and the chart would reproduce the values in the Portion of Total 1 column.

Copy to Other Spreadsheets or Programs

One of the greatest features of Windows or the Macintosh system is the clipboard. By copying any element of one document to the clipboard, you can paste it into another document of the same program or any other program that accepts clipboard information. For instance, you might want to include a chart in a word processor and print it. The word processor has much greater ability to present information in a readable fashion, though with effort you could accomplish similar results from the spreadsheet program.

Exercise 14

Select the range of cells from A5 to B7 and choose the command to make a pie chart or graph. Copy it to a word processing program and print it (File, Print). Save the spreadsheet.

Application Exercises

1. American Ginseng has three candidates for salesperson of the year: Alan, Patricia, and Diane. Prepare a spreadsheet that shows what each has contributed over the last three quarters. Include a heading, the date, totals and averages, and a chart comparing their performance.

   Table 1-3

	First Quarter	Second Quarter	Third Quarter
Alan	$111,258.97	$212,819.75	$213,567.35
Patricia	$211,798.83	$123,265.23	$221,058.37
Diane	$196,524.55	$176,985.28	$187,577.65

2. Add the sales figures from the fourth quarter to the spreadsheet you created in Application Exercise 1: Alan, $196,459.82; Patricia, $195,035.68; Diane, $188,657.98. What portion of the whole company's sales of $4,738,685.86 did each candidate contribute? (***Tip:*** *Enter the company total in a cell and use it as an absolute reference.*)

2. Checkbook Ledgers and Reconciliation

This chapter introduces a common procedure for both individuals and businesses: maintaining and balancing a checkbook. By hand and calculator, this task challenges even the best of accounting wizards. With a spreadsheet, the task is greatly simplified and much more easily accomplished. Its major advantage lies in the ease of correction and review of data. It also shows how time spent in preparing a spreadsheet reduces the work needed in the future because once you set up a bank reconciliation template, balancing subsequent statements is a snap.

Objectives

By working through and completing the exercises below, you will be able to:

- Prepare the labels for a checkbook ledger template.
- Input data, formulas and functions into a checkbook ledger template.
- Prepare the labels and reference cells for a checkbook reconciliation template.
- Input the formulas and functions for a checkbook reconciliation template.
- Make a template from a spreadsheet.

Checkbook Data Entry

The data needed for a checkbook ledger include its labels, descriptive text, and several kinds of numbers.

Checkbook Labels

The labels for a checkbook ledger are: Date, Check Number, Transaction Description (Payee, Source, or Purpose), Payment (Debit), Cleared?, Deposit (Credit), Fee, and Balance. Most of these are self explanatory. The transaction description column usually holds the payee's name for a payment and the payer's name for a deposit.

A payment is also called a debit while a deposit is also called a credit. Checkbook debits represent money leaving the account whereas credits represent money entering the account.

The Cleared? column lets you record whether or not the bank has processed a payment or deposit. This may be a single "Y" for yes and blank for no. The actual label does not need to be this long so we will use the question mark.

You will add two columns labeled Outstanding Payments and Outstanding Deposits. These will facilitate the reconciliation process. Reconciliation is most commonly referred to as Balancing a Checkbook. You will do this step in a separate column outside of the ledger area.

Exercise 1

To enter the checkbook labels:

1. Start a new spreadsheet file.
2. Select cell A1.
3. Enter **Checkbook for Marjorie Majors**.
4. Format cell A1 as bold.
5. Starting in cell A3 and moving across the row to B3, C3, and so on, enter: **Check No., Date, Transaction, Payment, ?, Fee, Deposit, Balance.** You should finish in cell H3.

6. Make cells A3 through H3 bold and centered in their cells.
7. Adjust all column widths to fit and wrap the text in cell A3.
8. Save the spreadsheet as CKLEDGER.

Your spreadsheet should resemble Figure 2-1.

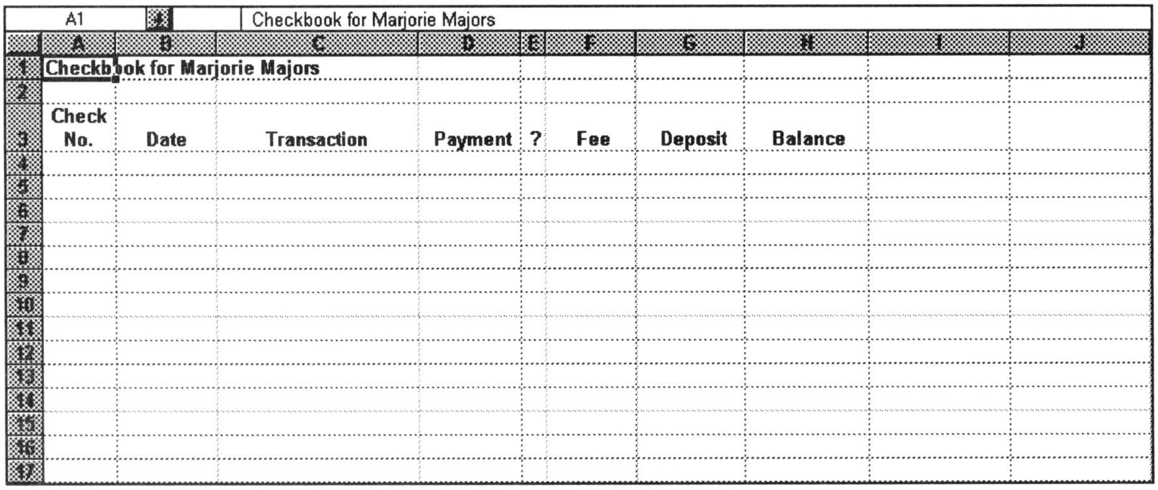

Figure 2-1: A checkbook ledger's heading labels.

Exercise 2

Complete the structure of the CKLEDGER spreadsheet by entering labels for the last two columns. Enter **Outstanding Payments** in cell I3 and **Outstanding Deposits** in cell J3. Center, wrap, and bold this text in the cell and widen the column as needed. Save the spreadsheet.

Your spreadsheet should look like Figure 2-2.

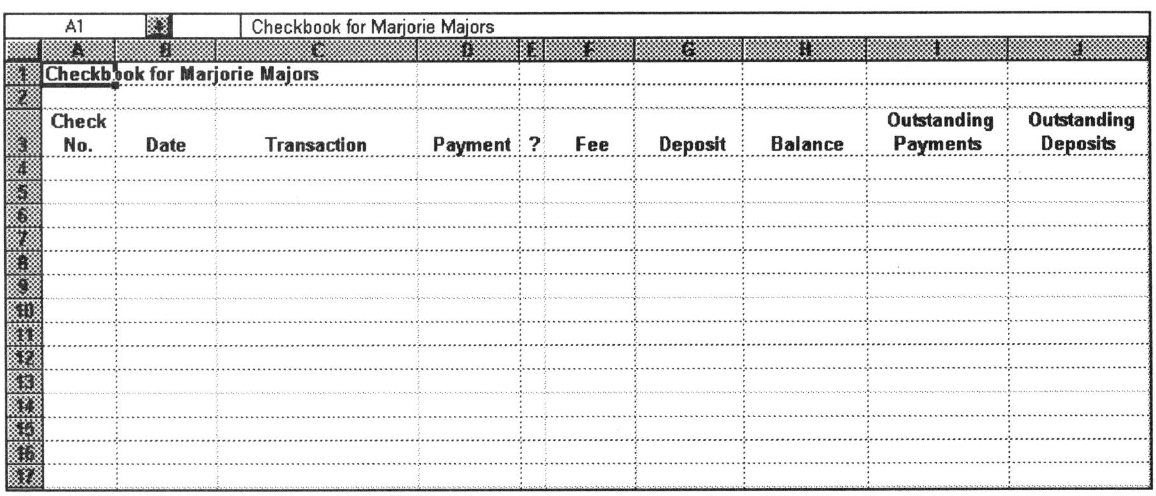

Figure 2-2: Preparing for reconciliation.

Checkbook Data

The checkbook ledger holds text data in the Transaction and Cleared? columns. In the Payment and Deposit columns you will enter numeric data formatted as currency. The Check Number column holds nominal numbers that should be ordinal. The Balance column holds a formula with relative references (no absolutes). The last two columns hold an IF function to list outstanding (not yet cleared) payments and deposits.

Exercise 3

Following Table 2-1 below, fill in the Check No, Date, Transaction, Payment, and Deposit data into Marjorie Major's checkbook ledger. Format columns D through J as currency with two decimal places. Begin in cell A5 and, when finished, save the spreadsheet.

Table 2-1

Number	Date	Description	Payment	Deposit
122	2/4	C. Salat, Consul	$ 60.00	
123	2/6	The Best Mag	$ 30.00	
124	2/14	Cost Giants	$ 96.00	
	2/16	School Service		$ 450.00
	2/16	Cash withdrawal	$ 50.00	
125	2/16	Visa payment	$ 250.00	
126	2/26	Home Hardware	$ 18.87	
127	3/4	Paper Supply	$ 77.92	

Your spreadsheet should look like Figure 2-3.

A1 | Checkbook for Marjorie Majors

	A	B	C	D	E	F	G	H	I	J
1	**Checkbook for Marjorie Majors**									
2										
3	**Check No.**	**Date**	**Transaction**	**Payment**	**?**	**Fee**	**Deposit**	**Balance**	**Outstanding Payments**	**Outstanding Deposits**
4										
5	122	2/4/96	C Salat, Consul	$ 60.00						
6	123	2/6/96	The Best Mag	$ 30.00						
7	124	2/14/96	Cost Giants	$ 96.00						
8		2/16/96	School Services				$ 450.00			
9		2/16/96	Cash	$ 50.00						
10	125	2/16/96	Visa payment	$ 250.00						
11	126	2/26/96	Home Hardware	$ 18.87						
12	127	3/4/96	Paper Supply	$ 77.92						
13										
14										
15										

Figure 2-3: Filling in the ledger.

Checkbook Formulas

The formulas needed for a checkbook ledger include the Balance column and totals at the bottom of the Outstanding Payments and Outstanding Deposits columns. The Balance makes good use of relative references to the Payments, Fees, and Deposits contents. It subtracts the payments and fees and adds the deposits. To plan for future checks, you should copy this formula down the column to cover a great many future checks.

Exercise 4

To create and copy the Balance formula:

1. Open the CKLEDGER file.
2. In cell H4 enter the beginning balance of $1,991.82.

3. In cell H5 enter =H4-D5-F5+G5.
4. Copy this formula from cell H5 down 30 rows to cell H34 (in real life you will want more rows ready for future use).
5. Save the spreadsheet.

Your spreadsheet should resemble Figure 2-4

A1 | Checkbook for Marjorie Majors

	A	B	C	D	E	F	G	H	I	J
1	**Checkbook for Marjorie Majors**									
2										
3	**Check No.**	**Date**	**Transaction**	**Payment**	**?**	**Fee**	**Deposit**	**Balance**	**Outstanding Payments**	**Outstanding Deposits**
4								$ 1,991.82		
5	122	2/4/96	C Salat, Consul	$ 60.00				$ 1,931.82		
6	123	2/6/96	The Best Mag	$ 30.00				$ 1,901.82		
7	124	2/14/96	Cost Giants	$ 96.00				$ 1,805.82		
8		2/16/96	School Services				$ 450.00	$ 2,255.82		
9		2/16/96	Cash	$ 50.00				$ 2,205.82		
10	125	2/16/96	Visa payment	$ 250.00				$ 1,955.82		
11	126	2/26/96	Home Hardware	$ 18.87				$ 1,936.95		
12	127	3/4/96	Paper Supply	$ 77.92				$ 1,859.03		
13								$ 1,859.03		
14								$ 1,859.03		
15								$ 1,859.03		

Figure 2-4: The balance formula provides instant balances with each ledger entry.

Checkbook Functions

The Outstanding Payments and Outstanding Deposits columns use a logical function. Their content depends on whether or not the contents in the Cleared? column is blank (no entry) or not blank (has data entered). If you specify that the entry must be "Y", you can be sure any other entry will be ignored.

Both columns will also need the Sum function used in their Totals row.

Exercise 5

To enter the logical formula for Outstanding Payments:

1. In cell I5 enter =IF(E5="",D5,"") or =IF(E5="Y",D5,"").
2. Copy this formula down to I34.
3. Save the spreadsheet.

Exercise 6

Enter the IF function formula into cell J5 to show the outstanding deposits. Save the spreadsheet.

Your spreadsheet should look like Figure 2-5.

A1 | Checkbook for Marjorie Majors

	A	B	C	D	E	F	G	H	I	J
1	**Checkbook for Marjorie Majors**									
2										
3	**Check No.**	**Date**	**Transaction**	**Payment**	**?**	**Fee**	**Deposit**	**Balance**	**Outstanding Payments**	**Outstanding Deposits**
4								$ 1,991.82		
5	122	2/4/96	C Salat, Consul	$ 60.00				$ 1,931.82	$ 60.00	$ -
6	123	2/6/96	The Best Mag	$ 30.00				$ 1,901.82	$ 30.00	$ -
7	124	2/14/96	Cost Giants	$ 96.00				$ 1,805.82	$ 96.00	$ -
8		2/16/96	School Services				$ 450.00	$ 2,255.82	$ -	$ 450.00
9		2/16/96	Cash	$ 50.00				$ 2,205.82	$ 50.00	$ -
10	125	2/16/96	Visa payment	$ 250.00				$ 1,955.82	$ 250.00	$ -
11	126	2/26/96	Home Hardware	$ 18.87				$ 1,936.95	$ 18.87	$ -
12	127	3/4/96	Paper Supply	$ 77.92				$ 1,859.03	$ 77.92	$ -
13								$ 1,859.03	$ -	$ -
14								$ 1,859.03	$ -	$ -
15								$ 1,859.03	$ -	$ -

Figure 2-5: The IF function prepares the outstanding payments and deposits for the reconciliation process.

Prepare the Reconciliation Process

The elements for creating the reconciliation portion of your checkbook ledger involves just seven labels, one number, two functions, and the rest are all simple formulas. However, getting that data involves the use of named ranges and some careful work comparing the ledger entries against the statement and fixing discrepancies. Usually, when the two disagree, the bank is correct, but at times the bank does make a mistake.

In completing the following exercises refer to Figure 2-6.

STATEMENT OF ACCOUNT

Prescott Credit Union
P.O. Box 6104 • XXXXXXX, XX XXXXXX • (XXX) XXX-XXXX

Marjorie Majors
1118 Montgomery Lane
XXXXXXX, XX XXXXXX

STATEMENT PERIOD
FROM TO
02-01-96 02-29-96
ACCOUNT NUMBER
407 467
PAGE 1

		PREVIOUS BALANCE		1991.82
2-29-96 TOTAL	2 DEPOSITS		1965.83	
2-29-96 TOTAL	1 WITHDRAWALS		2020.7	
2-29-96 TOTAL	0 OVERDRAFTS		00.0	
		ENDING BALANCE		1936.95

DATE	CHECK#	AMOUNT	DATE	CHECK#	AMOUNT	DATE	CHECK#	AMOUNT
2-13-96	122	60.00	2-16-96	124	96.00	2-28-96	126	18.87
2-20-96	123	30.00	2-26-96	125	250.00			

*BREAK IN NUMBER SEQUENCE

* * * DEPOSITS * * *
2/16/96 DEP FROM CHECKS RECEIVED 450.00

* * * WITHDRAWALS * * *
2/16/96 CASH DISBURSED 50.00

Figure 2-6: Marjorie Major's bank statement

Reconciliation Labels and Numbers

The labels needed for the reconciliation portion of your spreadsheet can be found on the back of most bank statements. You can place it anywhere on the spreadsheet but you must be careful not to place it in line with the ledger's columns since the column widths will differ. It is best to keep it away from the ledger's rows as well just in case a row gets inserted and potentially changes your formulas for the worst. In the example below, you will place the reconciliation portion below row 105 and to the right of column I.

Exercise 7

To enter the labels and the single number needed for the reconciliation portion of your spreadsheet:

1. Open the CKLEDGER spreadsheet.
2. Starting in cell L4 and moving down the column to L5, L6, and so on, enter **Ending Balance Shown on Statement, Total Outstanding Deposits, Sub-total, Total Outstanding Payments, Adjusted Ending Balance**. You should end in cell L8.
3. Select the L column and format its cells as bold, and make the width to fit.
4. In cell M4, enter the statement balance from Figure 2.6.
5. Format column M as currency with two decimal places and make its width to fit.
6. Save the spreadsheet.

The reconciliation portion of your spreadsheet should look like Figure 2-7.

M4 | 1936.95

	F	G	H	I	J	K	L	M
1								
2								
3	Fee	Deposit	Balance	Outstanding Payments	Outstanding Deposits			
4			$ 1,991.82				Ending Balance Shown on Statement	$ 1,936.95
5			$ 1,931.82	$ 60.00	$ -		Total Outstanding Deposits	
6			$ 1,901.82	$ 30.00	$ -		Sub-total	
7			$ 1,805.82	$ 96.00	$ -		Total Outstanding Payments	
8		$ 450.00	$ 2,255.82	$ -	$ 450.00		Adjusted Ending Balance	
9			$ 2,205.82	$ 50.00	$ -			
10			$ 1,955.82	$ 250.00	$ -			
11			$ 1,936.95	$ 18.87	$ -			
12			$ 1,859.03	$ 77.92	$ -			
13			$ 1,859.03	$ -	$ -			
14			$ 1,859.03	$ -	$ -			

Figure 2-7: The reconciliation labels.

Reconciliation Formulas

The formulas needed for reconciliation are simple references, the SUM function on a named range, addition, and subtraction.

Exercise 8

To enter the formulas for reconciling your checkbook ledger:

1. Select the range from J5 to J34 and name it the OUTDEP range.
2. In cell M5, enter = SUM(OUTDEP).
3. In cell M6, enter =M4+M5.
4. Select I5 to I105 and name it the OUTPAY range.
5. In cell M7, enter = SUM(OUTPAY).
6. In cell M8, enter =M6-M7.
7. Save the spreadsheet.

The reconciliation portion of your spreadsheet should look like Figure 2-8.

M8 | =M6-M7

	F	G	H	I	J	K	L	M
1								
2								
3	Fee	Deposit	Balance	Outstanding Payments	Outstanding Deposits			
4			$ 1,991.82				Ending Balance Shown on Statement	$ 1,936.95
5			$ 1,931.82	$ 60.00	$ -		Total Outstanding Deposits	$ 450.00
6			$ 1,901.82	$ 30.00	$ -		Sub-total	$ 2,386.95
7			$ 1,805.82	$ 96.00	$ -		Total Outstanding Payments	$ 582.79
8		$ 450.00	$ 2,255.82	$ -	$ 450.00		Adjusted Ending Balance	$ 1,804.16
9			$ 2,205.82	$ 50.00	$ -			
10			$ 1,955.82	$ 250.00	$ -			
11			$ 1,936.95	$ 18.87	$ -			
12			$ 1,859.03	$ 77.92	$ -			
13			$ 1,859.03	$ -	$ -			
14			$ 1,859.03	$ -	$ -			
15			$ 1,859.03	$ -	$ -			

Figure 2-8: With the reconciliation formulas in place, compare the adjusted ending balance with the ledger's balance.

Reconciliation Function

You may notice that the adjusted ending balance answer does not equal the ledger's balance. You must use the reconciliation process to make these two values equal. Before you start, however, enter one logical function to announce your success or failure to balance. To do so you will be introduced to the OFFSET and COUNT functions. The OFFSET function selects a cell in a range according to the number of cells down and to the right you designate. The COUNT function counts the number of cells in a given range. You must "nest" the COUNT function in the OFFSET function, which will be "nested" in an IF function.

> The OFFSET functions syntax is OFFSET(cell reference to start at, number of rows down, number of columns to the right, height of returned range, width of returned range). If you do not use the height and width parameters, the spreadsheet assumes the height and width will equal your selection.
>
> The COUNT syntax is: COUNT,(value 1, value 2, ...30). The values can be a variety of data types but only numbers are counted.

Exercise 9

To enter a function to help announce your success or failure:

1. Select the range from H4 to H34 and name it BALANCED.
2. In cell M10, enter =IF(M8=OFFSET(H4,COUNT(BALANCED)-1,0),"BALANCED","NOT!").
3. Save the spreadsheet.

Cell M10 should read NOT!

> *Hint:* To see what the difference is, in place of the above statement, enter: =IF(M8=OFFSET(H4,COUNT(BALANCED)-1,0),"BALANCED",M8-OFFSET(H4,COUNT(BALANCED)-1,0)).

Reconciliation Process

Now the real work begins. By comparing the statement against the ledger, entering missing data, marking cleared payments and deposits, and entering fee data, you will reconcile the differences between your checkbook ledger and your bank statement.

Exercise 10

To reconcile the bank statement and ledger, follow the steps below. Since you do not have the cleared checks on hand, assume the bank read each check and deposit slip correctly and the statement has the correct information.

1. Compare each check entry against the statement's list of cleared checks (Figure 2.6 on page 17).
2. Enter "Y" in the Cleared? column for every check that appears on the statement.
3. Enter checks missing from the ledger but included on the statement.
4. Enter the monthly bank charge in the Payments column and put a Y in the cleared column.
5. Cell M10 should now read, "BALANCED," so save the spreadsheet.

Make Template

The final step in this process is to make the sample spreadsheet into a template you can use for future projects and for the application exercises below. To do so, you must first protect the entire spreadsheet and then unlock the data entry cells. In this way you protect the formulas and allow data entry.

Exercise 11

1. Open the CKLEDGER spreadsheet.
2. So you can enter new data in the future, select cells A4 to at least G34 and format them as unlocked (Format, Cells, Protection, Unlock, for example).
3. To prevent any alterations of its formulas, find the command for and protect the entire spreadsheet (Tools, Sheet, Protect Sheet, for example).
4. Select and format cell M4 as unlocked.
5. Delete the data in cell M4 and cell A1 to G34.
6. Save the spreadsheet as CHCKTMP.

Now, whenever you use this template, remember to use the Save As command immediately and name the working spreadsheet a different name. In this way your template will always be clean of data and ready for your input. Also remember to insert rows before you use all the unlocked cells. Inserting rows also includes the inserted rows in the named ranges.

Application Exercises

Complete the bank ledgers and reconcile the following examples using your CHCKTMP template.

1. Open the template spreadsheet and save it as the POPELGR spreadsheet. Complete and reconcile the Pope Animal Clinic's checkbook ledger and statement.

Number	Date	Description of Transaction	Debit	✓	Fee	Credit	Balance
							2571.26
235	4/20	Pet Supply Company	42.95				
Deposit	5/1	Customer Receipts				110.00	
236	5/1	K-mart	12.15				
237	5/1	Telephone Company	72.63				
Deposit	5/8	Customer Receipts				200.00	
238	5/10	Chickasaw Electric Co.	95.73				
239	5/15	Protein Technologies	117.28				
240	5/15	Rond M Drug Co.	92.69				
Deposit	5/20	Customer Receipts				525.00	
Deposit	5/25	Customer Receipts				200.00	

STATEMENT OF ACCOUNT

Community First Bank
2177 Germantown Rd. South * Germantown, Tennessee 38138 * (901) 555-2400

Pope Animal Clinic
5012 Winchester
Memphis, TN 38118

STATEMENT PERIOD
FROM 5-1-96 **TO** 5-30-96
ACCOUNT NUMBER 43-7432156
PAGE 1

		PREVIOUS BALANCE	2571.28
5-30-96 TOTAL	3 DEPOSITS	835.00	
5-30-96 TOTAL	1 WITHDRAWALS	228.46	
5-30-96 TOTAL	0 OVERDRAFTS	00.00	
	ENDING BALANCE	3177.80	

DATE	CHECK#	AMOUNT	DATE	CHECK#	AMOUNT	DATE	CHECK#	AMOUNT
5-1-96	235	42.95	5-7-96	237	72.63			
5-3-96	236	12.15	5-15-96	238	95.73			

*BREAK IN NUMBER SEQUENCE

* * * DEPOSITS * * *

5-1-96 DEP FROM CHECKS RECEIVED	110.00
5-8-96 DEP FROM CHECKS RECEIVED	200.00
5-20-96 DEP FROM CHECKS RECEIVED	525.00

* * * WITHDRAWALS * * *

5-30-96 SERVICE FEE	5.00

2. Open the template spreadsheet and save it as the TOMLEDGR spreadsheet. Complete Tom Deskin's check ledger and balance it using the statement that follows.

Number	Date	Description of Transaction	Debit	✓	Fee	Credit	Balance
							2472 86
1094	8/28	K-mart	42 37				
1095	8/28	Walgreen's	12 96				
Deposit	9/1	Payroll Schering-Plough				2401 32	
AW	9/1	Leader Federal	942 18				
AW	9/1	L G & W	217 17				
1096	9/1	Kroger	36 01				
1097	9/1	Texaco	178 13				
1098	9/1	Univ. of Memphis	458 60				
1099	9/5	GMAC Credit Corp	583 21				
1100	9/8	Visa	283 21				
1101	9/10	Radio Shack	189 37				

Number	Date	Description of Transaction	Debit		✓	Fee	Credit		Balance	
1102	9/10	Auto Zone	48	23						
Deposit	9/15	Payroll-Schering Plough					2401	32		
1103	9/15	Geoffrey Beane	71	16						
1104	9/14	Heaven Scent Flowers	12	75						
1105	9/20	Kroger	87	85						
ATM	9/20	Kirby Woods	60	00						
1106	9/21	Traveler's Insurance	1238	42						
1107	9/23	Nation's Bank-Savings	500	00						

STATEMENT OF ACCOUNT

Community First Bank
2177 Germantown Rd. South * Germantown, Tennessee 38138 * (901) 555-2400

Tom Deskin
1234 South Street
Germantown, TN 38138

STATEMENT PERIOD
FROM TO
9-1-96 9-30-96
ACCOUNT NUMBER
13-2882139
PAGE 1

			PREVIOUS BALANCE		2472.86
9-30-96	TOTAL	2 DEPOSITS		4812.76	
9-30-96	TOTAL	1 WITHDRAWALS		4685.04	
9-30-96	TOTAL	0 OVERDRAFTS		00.0	
			ENDING BALANCE		2600.58

DATE	CHECK#	AMOUNT	DATE	CHECK#	AMOUNT	DATE	CHECK#	AMOUNT
8-31-96	1094	42.37	9-10-96	1099	583.21	9-25-96	1106	1238.42
9-2-96	1095	12.96	9-16-96	1100	283.21	9-25-96	1107	500.00
9-5-96	1096	36.01	9-18-96	1102	48.23			
9-5-96	1097	178.13	9-21-96	1103	71.16			
9-5-96	1098	458.60	9-23-96	1104	12.75			

*BREAK IN NUMBER SEQUENCE

* * * DEPOSITS * * *

9-1-96 DEP FROM CHECKS RECEIVED	2401.32
9-15-96 DEP FROM CHECKS RECEIVED	2401.32
9/57/96 INTEREST EARNED	10.12

* * * WITHDRAWALS * * *

9-1-96 CASH DISBURSED	942.18
9-4-96 CASH DISBURSED	217.17
9-20-96 CASH DISBURSED	60.00
9-27-96 FDIC INSURANCE PREMIUM	0.64

3. Open the template spreadsheet and save it as the TREELEDG spreadsheet. Balance the checkbook and statement for Tree Top Landscape.

Number	Date	Description of Transaction	Debit	✓	Fee	Credit	Balance
							4782.96
716	7/1	Dabney Nursery	90.23				
717	7/3	Office Max	42.78				
Deposit	7/3	Louis Lechleiter				200.00	
Deposit	7/5	Tony Trim				175.00	
Deposit	7/9	Dale Crosby				50.00	
718	7/10	Texaco Gas	29.36				
719	7/10	Nation's Bank	338.00				
Deposit	7/10	Bobby Cornelius				300.00	
ATM	7/15	Withdrawal	80.00				
720	7/20	AT&T	30.92				
721	7/20	Visa	172.83				

STATEMENT OF ACCOUNT

Community First Bank

2177 Germantown Rd. South * Germantown, Tennessee 38138 * (901) 555-2400

Tree Top Landscape Service
31125 Forest Hill-Lane Rd
Collierville, TN 38017

STATEMENT PERIOD
FROM 7-1-96 TO 8-2-96
ACCOUNT NUMBER 25-39042
PAGE 1

		PREVIOUS BALANCE		4782.96
8-2-96 TOTAL	DEPOSITS		725.00	
8-2-96 TOTAL	WITHDRAWALS		532.46	
8-2-96 TOTAL	OVERDRAFTS		00.0	
		ENDING BALANCE		4975.50

DATE	CHECK#	AMOUNT	DATE	CHECK#	AMOUNT	DATE	CHECK#	AMOUNT
7-5-96	716	90.23	7-12-96	718	29.36	7-22-96	720	30.92
7-7-96	717	42.78	7-15-96	719	238.00			

*BREAK IN NUMBER SEQUENCE

* * * DEPOSITS * * *

7-3-96 DEP FROM CHECKS RECEIVED	200.00
7-5-96 DEP FROM CHECKS RECEIVED	175.00
7-9-96 DEP FROM CHECKS RECEIVED	50.00
7-15-96 DEP FROM CHECKS RECEIVED	300.00

* * * WITHDRAWALS * * *

7-20-96 CASH DISBURSED	80.00
7-25-96 CHECK ORDER	21.17

4. Open the template spreadsheet and save it as the ENRILEDG spreadsheet. Enrique Anglade's checkbook ledger is shown below. Complete it and reconcile it with the bank statement.

Number	Date	Description of Transaction	Debit		✓	Fee	Credit		Balance	
									609	10
Deposit	4/1	Payroll					850	00		
Deposit	4/3	Payroll - Bonus					800	00		
5374	4/3	First Union Mortgage Co	647	53						
5375	4/3	South Florida Utility	82	75						
5376	4/5	1st Fed Crdt Union	219	95						
5377	4/15	Banc Boston	510	48						
Deposit	4/15	Payroll					850	00		
5378	4/20	Northwest Airlines	403	21						
5379	4/26	Auto Zone	18	99						
ATM	5/4	Cordova Branch	100	00						

STATEMENT OF ACCOUNT

Community First Bank
2177 Germantown Rd. South * Germantown, Tennessee 38138 * (901) 555-2400

Enrique Anglade
1901 Jones Drive
Miami, FL

STATEMENT PERIOD
FROM 4-1-96 TO 4-30-96
ACCOUNT NUMBER
32-123-32
PAGE 1

	PREVIOUS BALANCE		1034.10
4-30-96 TOTAL	DEPOSITS	2500.00	
4-30-96 TOTAL	WITHDRAWALS	962.73	
4-30-96 TOTAL	OVERDRAFTS	00.0	
	ENDING BALANCE		2571.37

DATE	CHECK#	AMOUNT	DATE	CHECK#	AMOUNT	DATE	CHECK#	AMOUNT
4-5-96	5374	647.53	4-8-96	5376	219.95			
4-5-96	5375	82.75						

*BREAK IN NUMBER SEQUENCE

* * * DEPOSITS * * *

4-1-96 DEP FROM CHECKS RECEIVED	850.00
4-3-96 DEP FROM CHECKS RECEIVED	800.00
7-9-96 DEP FROM CHECKS RECEIVED	850.00

* * * WITHDRAWALS * * *

4-30-96 SERVICE FEE	12.50

5. Open the template spreadsheet and save it as the TRACLEDG spreadsheet. Tracie Burke's Apparel Shop has the following checkbook entries. Balance the checkbook against the statement.

Number	Date	Description of Transaction	Debit	✓	Fee	Credit	Balance
							700.81
8213	5/28	Land's End	647.93				
Deposit	6/1	Receipts				1830.00	
8214	6/1	Collice Management Co	490.00				
8215	6/3	Jinkins Wholesale	738.32				
Deposit	6/5	Receipts				2583.00	
8216	6/5	Minn Utility Co.	257.13				
8217	6/10	State of MN	416.83				
Deposit	6/15	Receipts				3800.00	
8218	6/15	VOID					
8219	6/15	Tracie Burke Salary	2000.00				
8220	6/20	Brown's Wholesale	3150.00				
Deposit	7/2	Receipts				1720.00	

STATEMENT OF ACCOUNT

Community First Bank
2177 Germantown Rd. South • Germantown, Tennessee 38138 • (901) 555-2400

Tracie Burke's Apparel Shop
1396 Mall of America
Minneapolis, MN

STATEMENT PERIOD
FROM 6-1-96 TO 6-30-96
ACCOUNT NUMBER
12-324134523
PAGE 1

		PREVIOUS BALANCE	700.81
6-30-96	TOTAL	DEPOSITS	8218.83
6-30-96	TOTAL	WITHDRAWALS	5433.91
6-30-96	TOTAL	OVERDRAFTS	00.0
		ENDING BALANCE	3485.73

DATE	CHECK#	AMOUNT	DATE	CHECK#	AMOUNT	DATE	CHECK#	AMOUNT
6-2-96	8213	647.93	6-5-96	8215	728.32	6-20-96	8219	3150.00
6-3-96	8214	490.00	6-12-96	8217	416.83			

*BREAK IN NUMBER SEQUENCE

• • • DEPOSITS • • •

6-1-96	DEP FROM CHECKS RECEIVED	1830.00
6-5-96	DEP FROM CHECKS RECEIVED	2583.00
6-15-96	DEP FROM CHECKS RECEIVED	3800.00
6-30-96	INTEREST EARNED	5.83

• • • WITHDRAWALS • • •

6-30-96	FDIC INSURANCE	0.83

3. Financial Statements and Ratios

Financial statements monitor the status of a business. A Balance Sheet summarizes the assets, liabilities, and equity of a business as of a given date. The information on an Income Statement shows sales, expenses, and the profit or loss during a given period of time. Both the Balance Sheet and Income Statement can be further analyzed vertically or horizontally to extract more information about the business's condition. Further, their data can be used to generate various ratios indicating more about the business and helping owners, investors, and potential lenders and buyers make decisions.

Objectives

By working through and completing the exercises below, you will be able to:

- Prepare a balance sheet spreadsheet.
- Perform a vertical analysis for a balance sheet.
- Set up a comparative balance sheet.
- Prepare an income statement.
- Perform a vertical analysis for an income statement.
- Set up a comparative income statement.
- Perform a horizontal analysis of an income statement.
- Prepare named ranges for easy referencing.
- Prepare five financial ratios.
- Prepare a summary of ratio data table.

Balance Sheet Preparation

The data needed for preparing a business' balance sheet includes its assets, liabilities, and equity. These vary somewhat from business to business, so balance sheets may look different for different businesses. You will be building a template to use with any organization's financial reports.

Assets include current assets such as cash, accounts receivable (money owed to the business), inventory, and the like. It also includes plant assets such as equipment, vehicles, and buildings.

Liabilities include current liabilities and long-term liabilities. Current liabilities include accounts payable (money the business owes to others), wages payable, and the like. Long-term liabilities include notes payable, and mortgages.

Equity or net worth is the difference between assets and liabilities. It is the amount of clear ownership and is also called capital.

Balance Sheet

The labels in a balance sheet separate the assets, liabilities, and equity in that order. They further break down the current assets and plant assets and then the current liabilities and long-term liabilities. These entries are values and simple formulas including the SUM function.

For the balance sheet to be in balance, its total assets must equal the total of liabilities and equity.

Exercise 1

Enter and format the balance sheet labels as seen in Figure 3-1 and save the spreadsheet as FINSTATE. Notice the title is centered across columns A and B. Also notice the blank rows above the Total rows.

This blank row will allow you to add rows for future balance sheet items without affecting your formulas. In this way you are building a balance sheet template you can use for the application problems and in your work. To enter the formula for today's date use =TODAY().

A1 | Sander's Woodworks

	A	B	C	D
1	**Sander's Woodworks**			
2	**Balance Sheet**			
3	=TODAY()			
4				
5		**1996**		
6		**Amount**		
7	**Assets**			
8	*Current Assets:*			
9	Cash			
10	Accounts receivable			
11	Merchandise inventory			
12				
13	Total current assets			
14				
15	*Plant Assets*			
16	Equipment			
17				
18	Total plant			
19				
20	*Total Assets*			
21				
22	**Liabilities**			
23	*Current Liabilities*			
24	Accounts payable			
25	Wages payable			
26				
27	Total current liabilities			
28				
29	*Long-term liabilities*			
30	Mortage note payable			
31				
32	Total long-term liabilities			
33				
34	*Total Liabilities*			
35				
36	**Equity**			
37	J.Sanders Captial			
38				
39	Total Capital			
40				
41	Total Liabilities and Owner's Equity			
42				

Figure 3-1: Enter the above labels to start the Sander's Woodworks balance sheet.

Exercise 2

Enter and format the 1996 amounts in Table 3-1 in their matching cells.

Table 3-1

	1996	1995
Cash	$ 2,184.00	$ 1,973.00
Accounts receivable	$ 4,308.00	$ 2,118.00
Merchandise Inventory	$17,317.00	$18,476.00
Equipment	$14,203.00	$18,591.00
Accounts payable	$ 1,647.00	$ 2,317.00
Wages payable	$ 894.00	$ 684.00
Mortgage note payable	$12,715.00	$15,286.00
J. Sanders Capital	$22,756.00	$22,871.00

Save the spreadsheet. Your spreadsheet should look like Figure 3-2.

A1 | Sander's Woodworks

	A	B	C	D
1	**Sander's Woodworks**			
2	**Balance Sheet**			
3	**=TODAY()**			
4				
5		**1996**		
6		**Amount**		
7	**Assets**			
8	*Current Assets:*			
9	Cash	$ 2,184.00		
10	Accounts receivable	$ 4,308.00		
11	Merchandise inventory	$17,317.00		
12				
13	Total current assets			
14				
15	*Plant Assets*			
16	Equipment	$14,203.00		
17				
18	Total plant			
19				
20	*Total Assets*			
21				
22	**Liabilities**			
23	*Current Liabilities*			
24	Accounts payable	$ 1,647.00		
25	Wages payable	$ 894.00		
26				
27	Total current liabilities			
28				
29	*Long-term liabilities*			
30	Mortage note payable	$12,715.00		
31				
32	Total long-term liabilities			
33				
34	*Total Liabilities*			
35				
36	**Equity**			
37	J.Sanders Captial	$22,756.00		
38				
39	Total Capital			
40				
41	Total Liabilities and Owner's Equity			
42				

Figure 3-2: A balance sheet with its values.

Exercise 3

Enter the formulas for the balance sheet and bold their respective cells.

1. Use the SUM function to total the current assets in cell B13. Note the extra line is empty but will play an important role in maintaining your template's formulas: =SUM(B9:B12).
2. Use the SUM function to total plant assets (though there is only one in this example you will use the template and need the formula when you insert rows for actual financial reports): =SUM(B16:B17).
3. Use the SUM function to total current liabilities: =SUM(B24:B26).
4. Use the SUM function to total long-term liabilities: =SUM(B30:B31).
5. Use the SUM function to total capital (again the template only calls for one entry but future balance sheets may have more than one source of capital): =SUM(B37:B38).
6. For Total Assets, add Total Current Assets and Total Plant assets: =B13+B18.

7. For Total Liabilities, add Total Current Liabilities and Total Long-term Liabilities: =B27+B32.
8. For Total Liabilities and Owner's Equity, add Total Liabilities and Total Capital: =B27+B39.
9. Save the spreadsheet.

Your spreadsheet should look like Figure 3-3 below.

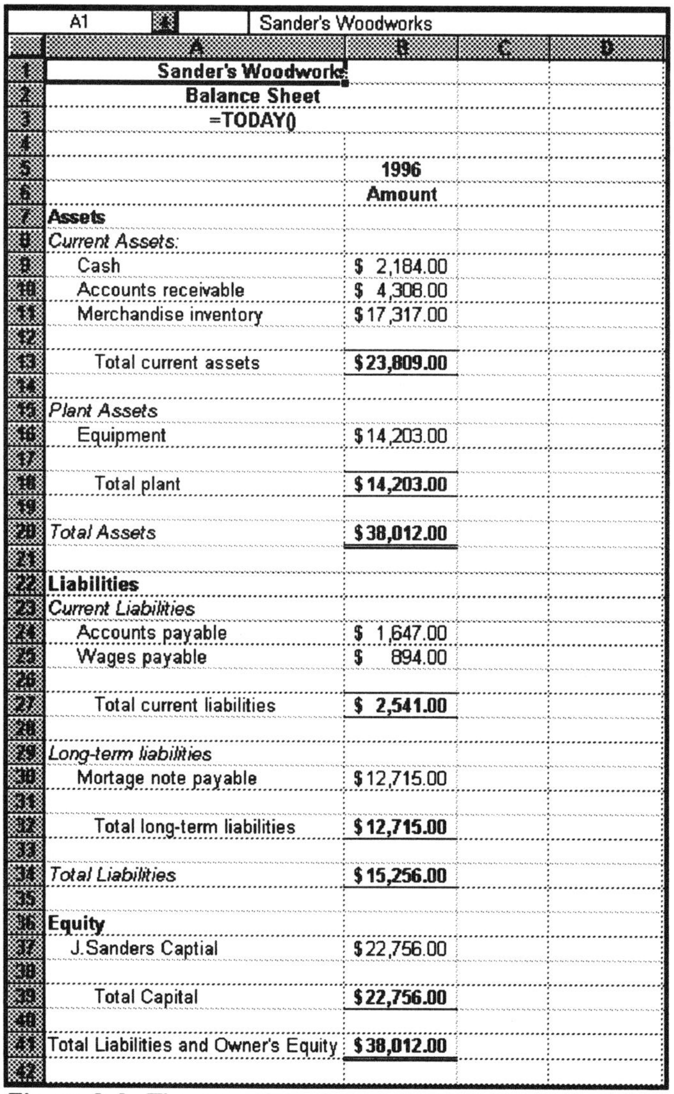

A1 | Sander's Woodworks

	A	B	C	D
1	Sander's Woodworks			
2	Balance Sheet			
3	=TODAY()			
4				
5		1996		
6		Amount		
7	**Assets**			
8	*Current Assets:*			
9	Cash	$ 2,184.00		
10	Accounts receivable	$ 4,308.00		
11	Merchandise inventory	$17,317.00		
12				
13	Total current assets	**$23,809.00**		
14				
15	*Plant Assets*			
16	Equipment	$14,203.00		
17				
18	Total plant	**$14,203.00**		
19				
20	*Total Assets*	**$38,012.00**		
21				
22	**Liabilities**			
23	*Current Liabilities*			
24	Accounts payable	$ 1,647.00		
25	Wages payable	$ 894.00		
26				
27	Total current liabilities	**$ 2,541.00**		
28				
29	*Long-term liabilities*			
30	Mortage note payable	$12,715.00		
31				
32	Total long-term liabilities	**$12,715.00**		
33				
34	*Total Liabilities*	**$15,256.00**		
35				
36	**Equity**			
37	J.Sanders Captial	$22,756.00		
38				
39	Total Capital	**$22,756.00**		
40				
41	Total Liabilities and Owner's Equity	**$38,012.00**		
42				

Figure 3-3: The complete Sander's Woodworks balance sheet.

Vertical Analysis of a Balance Sheet

The vertical analysis will show what portion each asset represents of totals assets. It also shows what portion each of the liabilities and the equity represent of Total Liabilities and Equity.

Exercise 4

To create the vertical analysis for Sander's Woodworks:

1. Open spreadsheet FINSTATE.

2. Format column C for percents with one decimal point.
3. Enter the label **Percent** into cell C6 centered and bolded.
4. In cell C9, enter the vertical analysis formula for assets using a partial absolute reference =B9/B$20.
5. Copy this formula down to C20.
6. Delete the formula from cells C14, C15, and C19.
7. In cell C24, enter the vertical analysis formula for liabilities using a partial absolute reference: =B24/B$41.
8. Copy this formula down to C41.
9. Delete the formula from cells C28, C29, C33, C35, C36, and C40.
10. Format the spreadsheet as you can see in Figure 3-4 below.
11. Save the spreadsheet.

Your spreadsheet should look like Figure 3-4.

A1 Sander's Woodworks

	A	B	C	D	E
1	**Sander's Woodworks**				
2	**Balance Sheet**				
3	**=TODAY()**				
4					
5		**1996**			
6		**Amount**	**Percent**		
7	**Assets**				
8	*Current Assets:*				
9	Cash	$ 2,184.00	5.7%		
10	Accounts receivable	$ 4,308.00	11.3%		
11	Merchandise inventory	$17,317.00	45.6%		
12					
13	Total current assets	**$23,809.00**	62.6%		
14					
15	*Plant Assets*				
16	Equipment	$14,203.00	37.4%		
17					
18	Total plant	**$14,203.00**	37.4%		
19					
20	*Total Assets*	**$38,012.00**	100.0%		
21					
22	**Liabilities**				
23	*Current Liabilities*				
24	Accounts payable	$ 1,647.00	4.3%		
25	Wages payable	$ 894.00	2.4%		
26					
27	Total current liabilities	**$ 2,541.00**	6.7%		
28					
29	*Long-term liabilities*				
30	Mortage note payable	$12,715.00	33.4%		
31					
32	Total long-term liabilities	**$12,715.00**	33.4%		
33					
34	*Total Liabilities*	**$15,256.00**	40.1%		
35					
36	**Equity**				
37	J.Sanders Captial	$22,756.00	59.9%		
38					
39	Total Capital	**$22,756.00**	59.9%		
40					
41	Total Liabilities and Owner's Equity	**$38,012.00**	100.0%		
42					

Figure 3-4: This vertical analysis now shows how much of the totals each amount represents.

Exercise 5

Open the FINSTATE spreadsheet and save it as the SANDERS spreadsheet. Copy the Balance Sheet to a word processing document and write a descriptive paragraph summarizing the results. For example, "Current assets account for 62.6% of the total assets with 45.6% (73% of current assets) of that invested in inventory." Illustrate the results of your ananysis with a pie graph made in the spreadsheet program such as the following:

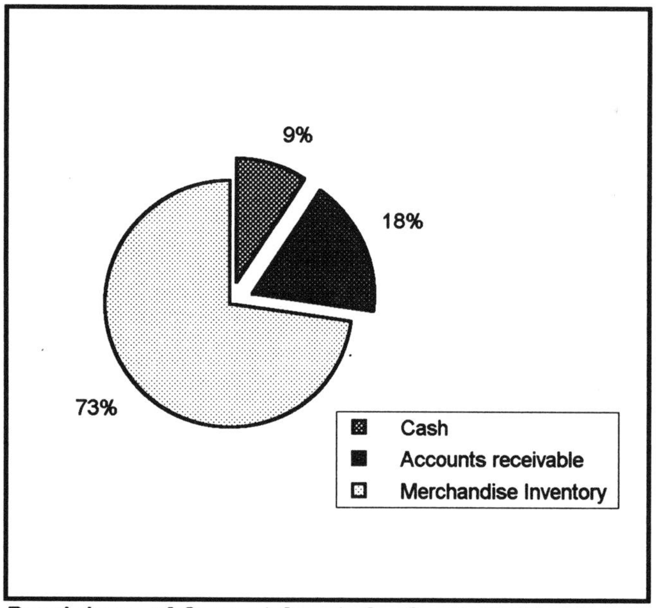

Breakdown of Current Assets for Sander's Woodworks 1996.

Comparative Balance Sheet

It is often advantageous to display a business' performance along with its performance from a previous and comparable time. For instance it is often helpful for planning a new year to show the current year's ending balance sheet along side the previous year's ending balance sheet. This can be easily done building on the FINSTATE spreadsheet and entering values from a previous date.

Exercise 6

To create a horozontal comparative analysis balance sheet:

1. Open spreadsheet FINSTATE.
2. Center **1996** across columns B and C and enter **1995** in D5 centered across D and E.
3. Select the range B9 to C41 and copy it to cell D9.
4. Replace the amounts in column D with the 1995 amounts from Table 3-1 on page 28.
5. Format the worksheet as seen in Figure 3-5 on page 33 (notice the title is centered across all the used columns).
6. Save the spreadsheet.

Your spreadsheet should resemble Figure 3-5.

A1 | Sander's Woodworks

Sander's Woodworks
Balance Sheet
=TODAY()

	1996 Amount	1996 Percent	1995 Amount	1995 Percent
Assets				
Current Assets:				
Cash	$ 2,184.00	5.7%	$ 1,973.00	4.8%
Accounts receivable	$ 4,308.00	11.3%	$ 2,118.00	5.1%
Merchandise inventory	$17,317.00	45.6%	$18,476.00	44.9%
Total current assets	**$23,809.00**	62.6%	**$22,567.00**	54.8%
Plant Assets				
Equipment	$14,203.00	37.4%	$18,591.00	45.2%
Total plant	**$14,203.00**	37.4%	**$18,591.00**	45.2%
Total Assets	**$38,012.00**	100.0%	**$41,158.00**	100.0%
Liabilities				
Current Liabilities				
Accounts payable	$ 1,647.00	4.3%	$ 2,317.00	5.6%
Wages payable	$ 894.00	2.4%	$ 684.00	1.7%
Total current liabilities	**$ 2,541.00**	6.7%	**$ 3,001.00**	7.3%
Long-term liabilities				
Mortage note payable	$12,715.00	33.4%	$15,286.00	37.1%
Total long-term liabilities	**$12,715.00**	33.4%	**$15,286.00**	37.1%
Total Liabilities	**$15,256.00**	40.1%	**$18,287.00**	44.4%
Equity				
J.Sanders Captial	$22,756.00	59.9%	$22,871.00	55.6%
Total Capital	**$22,756.00**	59.9%	**$22,871.00**	55.6%
Total Liabilities and Owner's Equity	**$38,012.00**	100.0%	**$41,158.00**	100.0%

Figure 3-5: Showing two years' analyses help businesses decide on what to do in coming months or years.

Exercise 7

Open and save spreadsheet FINSTATE as the SANDERS spreadsheet. Delete the extra rows and copy the comparative balance sheet to the end of the report you wrote in Exercise 5 and write a descriptive paragraph about the 1995 results as compared to the 1996 results. For example, "In 1995, Accounts Receivable accounted for only 5.1% of total assets while in 1996 this has increased to 11.3%." Create bar charts in the spreadsheet program and copy them to your report to show the differences between 1995 and 1996 totals such as:

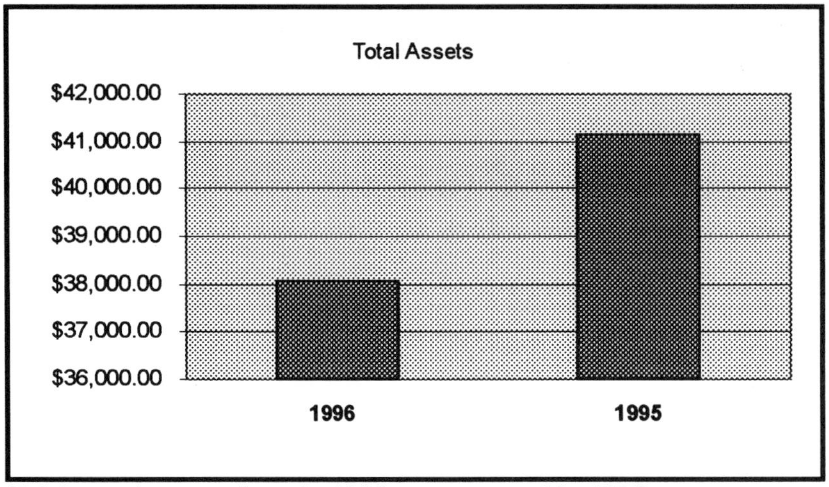

Total assets were down in 1996 as compared to 1995.

A horizontal analysis compares balance sheet results between two comparative times indicating an increase or decrease from the earlier date and later date.

Exercise 8

To create a horizontal analysis:

1. Open the FINSTATE spreadsheet.
2. In cell F5, enter **Increase** or **Decrease** (center across columns F and G).
3. Enter **Amount** and **Percent** in cells F6 and G6 receptively.
4. In cell F9, enter the formula to show the difference between the 1996 and 1995 amounts: =B9-D9.
5. Copy this formula down to F41 and then delete it from what should be empty cells (F14, F15, F19, F21, F22, F23, F28, F29, F33, F35, F36, and F40).
6. In cell G9, enter the formula showing the percent of increase or decrease between 1996 and 1995 that the 1995 amount represents: **=F9/D9**.
7. Format column G as a percent with one decimal place.
8. Copy the percent of increase formula down to G41 and then delete it from what should be empty cells (G14, G15, G19, G21, G22, G23, G28, G29, G33, G35, G36, and G40). Note the **#DIV0!** message in the blank "holding" cells. This means the formula attempts to divide by zero, which is not possible.

   Tip: *You can use an IF formula to eliminate the* **#DIV0!** *message: =IF(Increase or Decrease Amount=0,then "",otherwise Amount/1995 Amount), which in our example is* =IF(F9=0,"",F9/D9)*.*

9. Save the spreadsheet.

Your spreadsheet should look like Figure 3-6.

A1 | Sander's Woodworks

	1996 Amount	1996 Percent	1995 Amount	1995 Percent	Increase or Decrease Amount	Increase or Decrease Percent
	Sander's Woodworks					
	Balance Sheet					
	=TODAY()					
Assets						
Current Assets:						
Cash	$ 2,184.00	5.7%	$ 1,973.00	4.8%	$ 211.00	10.7%
Accounts receivable	$ 4,308.00	11.3%	$ 2,118.00	5.1%	$ 2,190.00	103.4%
Merchandise inventory	$17,317.00	45.6%	$18,476.00	44.9%	$(1,159.00)	-6.3%
Total current assets	**$23,809.00**	62.6%	**$22,567.00**	54.8%	$ 1,242.00	5.5%
Plant Assets						
Equipment	$14,203.00	37.4%	$18,591.00	45.2%	$(4,388.00)	-23.6%
Total plant	**$14,203.00**	37.4%	**$18,591.00**	45.2%	$(4,388.00)	-23.6%
Total Assets	**$38,012.00**	100.0%	**$41,158.00**	100.0%	$(3,146.00)	-7.6%
Liabilities						
Current Liabilities						
Accounts payable	$ 1,647.00	4.3%	$ 2,317.00	5.6%	$ (670.00)	-28.9%
Wages payable	$ 894.00	2.4%	$ 684.00	1.7%	$ 210.00	30.7%
Total current liabilities	**$ 2,541.00**	6.7%	**$ 3,001.00**	7.3%	$ (460.00)	-15.3%
Long-term liabilities						
Mortage note payable	$12,715.00	33.4%	$15,286.00	37.1%	$(2,571.00)	-16.8%
Total long-term liabilities	**$12,715.00**	33.4%	**$15,286.00**	37.1%	$(2,571.00)	-16.8%
Total Liabilities	**$15,256.00**	40.1%	**$18,287.00**	44.4%	$(3,031.00)	-16.6%
Equity						
J.Sanders Captial	$22,756.00	59.9%	$22,871.00	55.6%	$ (115.00)	-0.5%
Total Capital	**$22,756.00**	59.9%	**$22,871.00**	55.6%	$ (115.00)	-0.5%
Total Liabilities and Owner's Equity	**$38,012.00**	100.0%	**$41,158.00**	100.0%	$(3,146.00)	-7.6%

Figure 3-6: The horizontal analysis shows the amount of increase or decrease in account balances from one year to the next.

Exercise 9

Open the FINSTATE spreadsheet and save it as the SANDERS spreadsheet. Remove the extra rows and copy the whole balance sheet with its vertical, comparative, and horizontal analyses to the end of the report you created in Exercises 5 and 7. Write a descriptive paragraph highlighting the differences between 1996 and 1995 amounts. For example, "While total current assets have increased by 5.5%, total assets have actually decreased by 7.6%, which may be due the sale of outmoded equipment in 1996." Support your statements with illustrative charts.

Income Statement Preparation

Businesses need to know what their net income is over a given time period. The income statement provides this information.

Income Statement

The income statement has five basic parts: Net Sales, Cost of Goods Sold, Gross Profit, Operating Expenses, and Net Income. Each is a formula and will be created in the statement using formulas.

Exercise 10

Enter the Income Statement labels into spreadsheet FINSTATE. Enter and format the text as seen in Figure 3-7 and save the spreadsheet. Notice that we have moved down to give the Balance Sheet room to grow and moved over so as not to affect the column widths of the balance sheet. Again, notice the empty row above the Total Operating Expenses row to allow for expansion in future examples using this template.

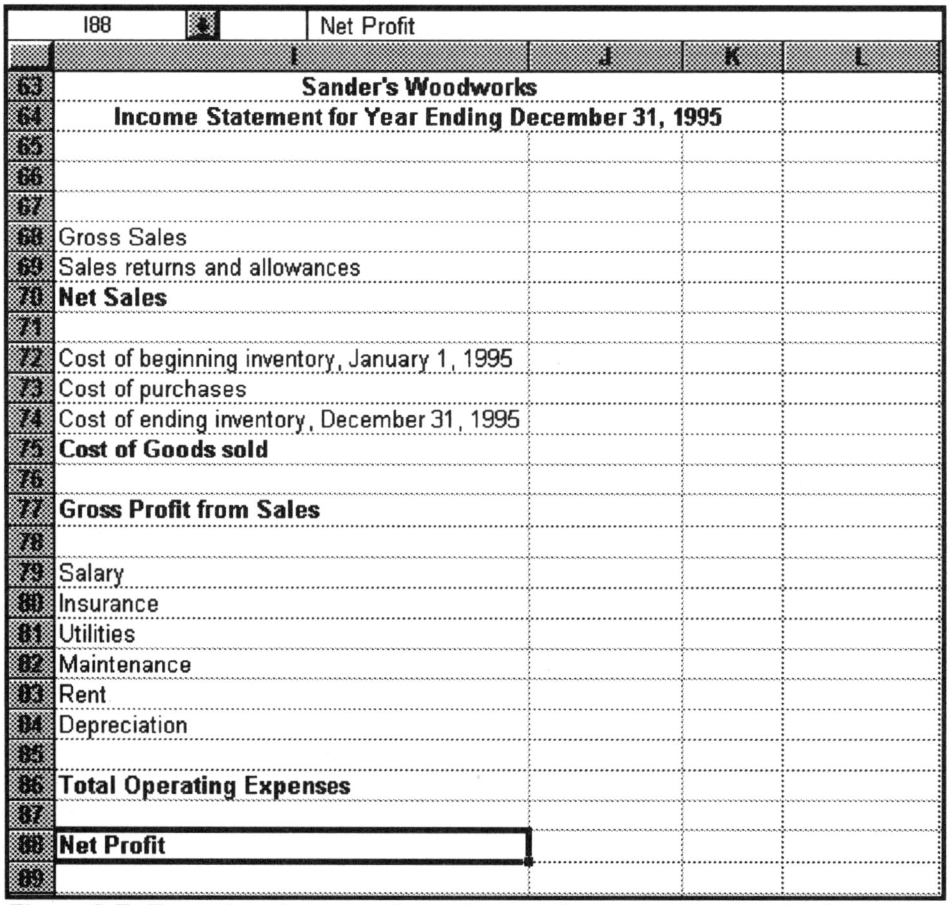

Figure 3-7: Enter the above labels to start the Sander's Woodwork's income statement.

Exercise 11

Enter the 1996 amounts from Table 3-2 in their matching cells and save the spreadsheet.

Table 3-2

	1996	**1995**
Gross Sales	$246,891.00	$228,756.00
Sales returns and allowances	$ 7,835.00	$ 6,954.00
Cost of beginning inventory	$ 8,247.00	$ 7,965.00
Cost of purchases	$148,542.00	$142,769.00
Cost of ending inventory	$ 9,583.00	$ 8,247.00
Salary	$ 18,500.00	$ 17,400.00
Insurance	$ 5,700.00	$ 5,400.00
Utilities	$ 1,900.00	$ 1,850.00
Maintenance	$ 280.00	$ 265.00
Rent	$ 6,000.00	$ 6,000.00
Depreciation	$ 1,500.00	$ 1,450.00

Exercise 12

In the FINSTATE spreadsheet, enter the formulas for the income statement in their appropriate cells:

1. Net Sales = gross sales - returns and allowances: =J68-J69.
2. Cost of goods sold = cost of beginning inventory + cost of purchases - cost of ending inventory: =J72+J73-J74.
3. Gross profit from sales = net sales - cost of goods sold: =J70-J75.
4. Use the SUM function to find the Total operating expenses: =SUM(J79:J85). Remember to include the extra row.
5. Net profit = gross profit - operating expenses: =J77-J86
6. Save the spreadsheet.

Your spreadsheet should look like Figure 3-8 below.

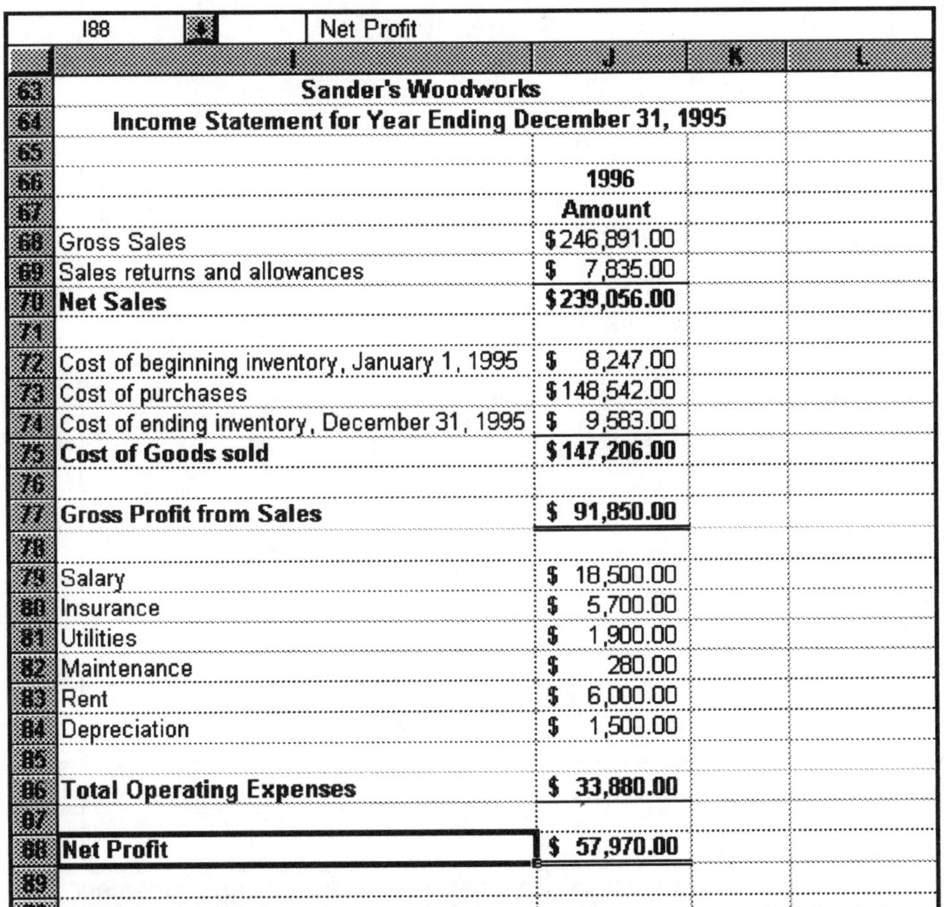

I88 | Net Profit

	I	J	K	L
63	**Sander's Woodworks**			
64	**Income Statement for Year Ending December 31, 1995**			
65				
66		**1996**		
67		**Amount**		
68	Gross Sales	$246,891.00		
69	Sales returns and allowances	$ 7,835.00		
70	**Net Sales**	**$239,056.00**		
71				
72	Cost of beginning inventory, January 1, 1995	$ 8,247.00		
73	Cost of purchases	$148,542.00		
74	Cost of ending inventory, December 31, 1995	$ 9,583.00		
75	**Cost of Goods sold**	**$147,206.00**		
76				
77	**Gross Profit from Sales**	**$ 91,850.00**		
78				
79	Salary	$ 18,500.00		
80	Insurance	$ 5,700.00		
81	Utilities	$ 1,900.00		
82	Maintenance	$ 280.00		
83	Rent	$ 6,000.00		
84	Depreciation	$ 1,500.00		
85				
86	**Total Operating Expenses**	**$ 33,880.00**		
87				
88	**Net Profit**	**$ 57,970.00**		
89				
90				

Figure 3-8: The complete Sander's Woodworks income statement.

Vertical Analysis of an Income Statement

A common vertical analysis is a percent of sales analysis, which we will do here. Another useful kind would be a percent of expenses analysis. The percent of sales vertical analysis of an income statement shows owners and investors what portion of the net sales each item on the statement represents.

Exercise 13

To create a percent of sales vertical analysis for the Sander's Woodworks income statement:

1. Open the FINSTATE spreadsheet.

2. Enter **Percent** in cell K67.
3. Bold and center the Amount and Percent labels and center 1996 across columns J66 and K66.
4. Enter the formula to find the percent of sales in cell K68: **=J68/J$70**.
5. Copy this formula from K68 to K88 and remove it from K71, K76, K78, K85, and K87.
6. Save the spreadsheet.

Your spreadsheet should resemble Figure 3-9.

I88 | Net Profit

	I	J	K	L
63	**Sander's Woodworks**			
64	**Income Statement for Year Ending December 31, 1995**			
65				
66		**1996**		
67		**Amount**	**Percent**	
68	Gross Sales	$246,891.00	103.3%	
69	Sales returns and allowances	$ 7,835.00	3.3%	
70	**Net Sales**	**$239,056.00**	100.0%	
71				
72	Cost of beginning inventory, January 1, 1995	$ 8,247.00	3.4%	
73	Cost of purchases	$148,542.00	62.1%	
74	Cost of ending inventory, December 31, 1995	$ 9,583.00	4.0%	
75	**Cost of Goods sold**	**$147,206.00**	61.6%	
76				
77	**Gross Profit from Sales**	**$ 91,850.00**	38.4%	
78				
79	Salary	$ 18,500.00	7.7%	
80	Insurance	$ 5,700.00	2.4%	
81	Utilities	$ 1,900.00	0.8%	
82	Maintenance	$ 280.00	0.1%	
83	Rent	$ 6,000.00	2.5%	
84	Depreciation	$ 1,500.00	0.6%	
85			0.0%	
86	**Total Operating Expenses**	**$ 33,880.00**	14.2%	
87				
88	**Net Profit**	**$ 57,970.00**	24.2%	
89				
90				

Figure 3-9: The complete vertical analysis for the Sander's Woodworks income statement.

Exercise 14

Open the FINSTATE spreadsheet template and save it as the SANDERS spreadsheet (if you want to save your graphs from above, save it as a version of the SANDERS spreadsheet such as SANDERS2). Copy the percent of sales vertical analysis income statement to your report document and write a paragraph highlighting the results. For example, "Cost of goods sold covers 61.6% of the sales." Create pie charts showing the ways to analysis the data such as:

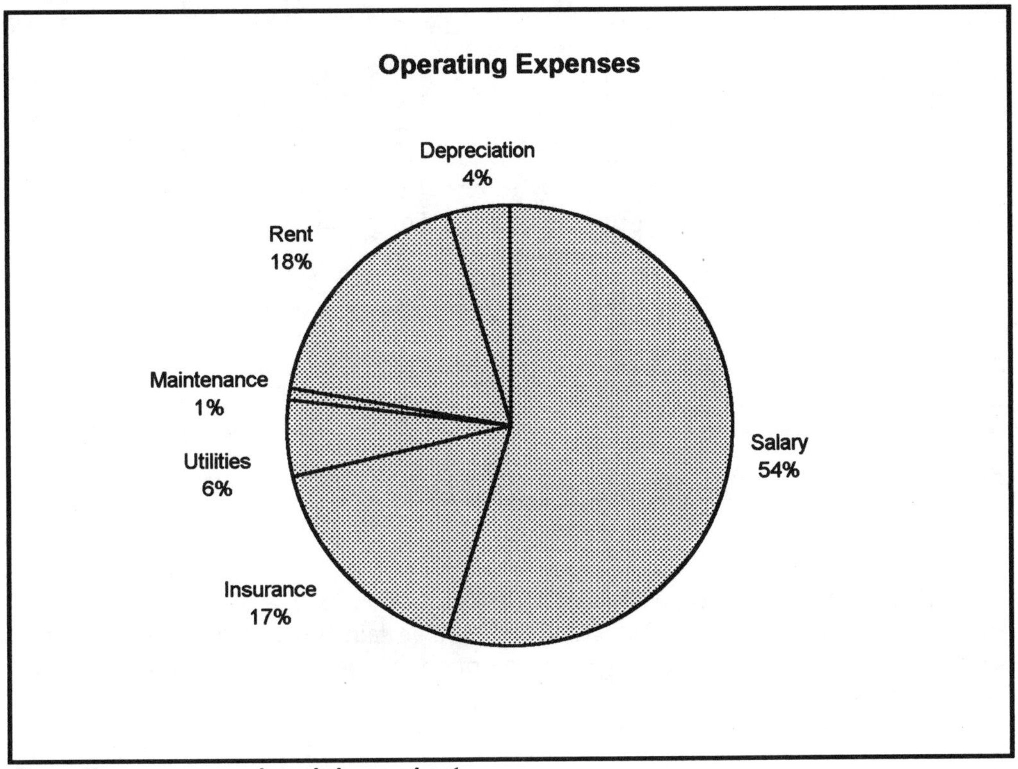

Operating expenses breakdown chart.

Comparative Income Statement

The comparative analysis of an income statement over different time periods also provides valuable information to business owners and investors.

Exercise 15

To create a comparative horizontal analysis for the Sander's Woodworks income statement:

1. Open the FINSTATE spreadsheet.
2. Add the **1995** label
3. Copy the range J68:K88 to L68.
4. Enter the data for 1995 from Table 3-2 on page 36.
5. Format and save the spreadsheet.

You spreadsheet should look like Figure 3-10 on page 40.

I88 Net Profit

	I	J	K	L	M
63	**Sander's Woodworks**				
64	**Income Statement for Year Ending December 31, 1995**				
65					
66		**1996**		**1995**	
67		**Amount**	**Percent**	**Amount**	**Percent**
68	Gross Sales	$246,891.00	103.3%	$228,756.00	103.1%
69	Sales returns and allowances	$ 7,835.00	3.3%	$ 6,954.00	3.1%
70	**Net Sales**	**$239,056.00**	100.0%	**$221,802.00**	100.0%
71					
72	Cost of beginning inventory, January 1, 1995	$ 8,247.00	3.4%	$ 7,965.00	3.6%
73	Cost of purchases	$148,542.00	62.1%	$142,769.00	64.4%
74	Cost of ending inventory, December 31, 1995	$ 9,583.00	4.0%	$ 9,427.00	4.3%
75	**Cost of Goods sold**	**$147,206.00**	61.6%	**$141,307.00**	63.7%
76					
77	**Gross Profit from Sales**	**$ 91,850.00**	38.4%	**$ 80,495.00**	36.3%
78					
79	Salary	$ 18,500.00	7.7%	$ 17,400.00	7.8%
80	Insurance	$ 5,700.00	2.4%	$ 5,400.00	2.4%
81	Utilities	$ 1,900.00	0.8%	$ 1,850.00	0.8%
82	Maintenance	$ 280.00	0.1%	$ 265.00	0.1%
83	Rent	$ 6,000.00	2.5%	$ 6,000.00	2.7%
84	Depreciation	$ 1,500.00	0.6%	$ 1,450.00	0.7%
85			0.0%		0.0%
86	**Total Operating Expenses**	**$ 33,880.00**	14.2%	**$ 32,365.00**	14.6%
87					
88	**Net Profit**	**$ 57,970.00**	24.2%	**$ 48,130.00**	21.7%
89					
90					

Figure 3-10: The Sander's Woodworks horizontal income statement.

Exercise 16

Open and save the FINSTATE spreadsheet as the SANDERS spreadsheet or a version of it. Copy the analysis to your report document and write a paragraph highlighting salient points. For example, "In 1996, the 2% reduction in cost of goods sold created a better profit margin." Support your work with bar charts such as:

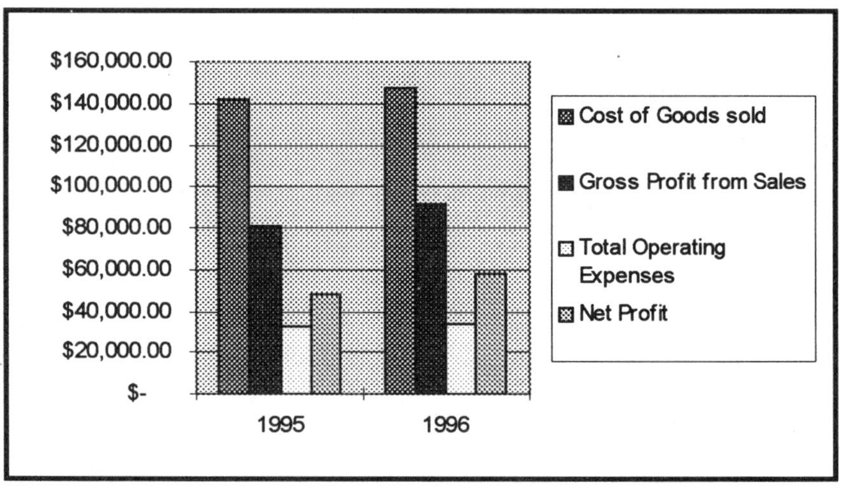

The Sander's Woodworks income statement comparisons for 1995 and 1996.

Horizontal Analysis of an Income Statement

The horizontal analysis of an income statement shows the amount and percent of increase or decrease between two time periods.

Exercise 17

To create the horizontal analysis for Sander's Woodworks:

1. Open the FINSTATE file.
2. In cell E6 enter **Increase or Decrease** and format it as bold and centered across N66 and O66.
3. In cells N67 and O67 enter **Amount** and **Percent** bolded and centered.
4. In cell N68, enter the formula showing the difference between 1996 and 1995 amounts: =J68-L68.
5. In cell O68, enter the formula showing what percent of increase or decrease the amount in N67 represents: =N68/L68.
6. Copy these formulas down to cells N88 and O88 and remove it from the holding cells.
7. Format all the entries as shown in Figure 3-11 and save the spreadsheet.

Your spreadsheet should look like Figure 3-11.

I88 | Net Profit

	I	J	K	L	M	N	O
63	**Sander's Woodworks**						
64	**Income Statement for Year Ending December 31, 1995**						
65							
66		**1996**		**1995**		**Increase or Decrease**	
67		**Amount**	**Percent**	**Amount**	**Percent**	**Amount**	**Percent**
68	Gross Sales	$246,891.00	103.3%	$228,756.00	103.1%	$18,135.00	7.9%
69	Sales returns and allowances	$ 7,835.00	3.3%	$ 6,954.00	3.1%	$ 881.00	12.7%
70	**Net Sales**	**$239,056.00**	100.0%	**$221,802.00**	100.0%	$17,254.00	7.8%
71							
72	Cost of beginning inventory, January 1, 1995	$ 8,247.00	3.4%	$ 7,965.00	3.6%	$ 282.00	3.5%
73	Cost of purchases	$148,542.00	62.1%	$142,769.00	64.4%	$ 5,773.00	4.0%
74	Cost of ending inventory, December 31, 1995	$ 9,583.00	4.0%	$ 9,427.00	4.3%	$ 156.00	1.7%
75	**Cost of Goods sold**	**$147,206.00**	61.6%	**$141,307.00**	63.7%	$ 5,899.00	4.2%
76							
77	**Gross Profit from Sales**	**$ 91,850.00**	38.4%	**$ 80,495.00**	36.3%	$11,355.00	14.1%
78							
79	Salary	$ 18,500.00	7.7%	$ 17,400.00	7.8%	$ 1,100.00	6.3%
80	Insurance	$ 5,700.00	2.4%	$ 5,400.00	2.4%	$ 300.00	5.6%
81	Utilities	$ 1,900.00	0.8%	$ 1,850.00	0.8%	$ 50.00	2.7%
82	Maintenance	$ 280.00	0.1%	$ 265.00	0.1%	$ 15.00	5.7%
83	Rent	$ 6,000.00	2.5%	$ 6,000.00	2.7%	$ -	0.0%
84	Depreciation	$ 1,500.00	0.6%	$ 1,450.00	0.7%	$ 50.00	3.4%
85							
86	**Total Operating Expenses**	**$ 33,880.00**	14.2%	**$ 32,365.00**	14.6%	$ 1,515.00	4.7%
87							
88	**Net Profit**	**$ 57,970.00**	24.2%	**$ 48,130.00**	21.7%	$ 9,840.00	20.4%
89							

Figure 3-11: The Sander's Woodworks horizontal analysis income statement.

Exercise 18

It would be nice to have more years in the horizontal analysis to make a more interesting line graph and chart but save the FINSTATE spreadsheet as the SANDERS spreadsheet or a version of it. Add it to your report document with supportive text and charts.

Name Ranges

To support the upcoming work in which you will reference several cells throughout the spreadsheet, it will be helpful to name the cells. Otherwise you will be scrolling all over the spreadsheet trying to find cell addresses long forgotten.

Exercise 19

To create the named ranges needed for your ratio formulas:

1. Open the FINSTATE spreadsheet.
2. In the balance sheet:
    - Name the Merchandise Inventory cell in column B, **INVENTORY**.
    - Name the Total Current Assets cell in column B, **TOTALCURRENTASSETS**.
    - Name the Total Current Liabilities cell in column B, **TOTALCURRENTLIA**.
    - Name the Total Assets cell in column B, **TOTALASSETS**.
    - Name the Total Liabilities cell in column B, **TOTALLIA**.
3. In the income statement:
    - Name the Cost of Goods Sold cell in column J, **COGS**.
    - Name the Total Operating Expenses cell in column J, **EXPENSES**.
    - Name the Net Sales cell in column J, **NETSALES**.

To support going to, selecting, and printing ranges:

1. In the balance sheet:
    - Name cell A1 as **NAME**.
    - Name cell A3 as **MONTHYR**.
    - Name the range covering the whole balance sheet as **BALANCE**.
2. Name the range covering the whole income statement as **INCOME**.

Financial Ratios Preparation

Financial ratios give business owners a different perspective on their financial condition and allow them to compare the business' financial health with industry standards. Different values indicate different financial states for a business. Review Table 3-3 and then complete the exercises to learn more about these handy guides. The exercises have you create a Financial Ratio worksheet and a report form for each type of ratio.

Table 3-3

Ratio	Formula	Analysis
Current Ratio (RatioCR)	=TotalCurrentAssets/TotalCurrentLia	=IF(RatioCR<1,"Debts greater than assets.",IF (RatioCR=1,"Assets equal to Debts.",IF (RatioCR>=2,"Desirable!","Assets greater than debts.")))
Acid-test ratio (RatioAT)	=QuickAssets/TotalLia	=IF(RatioAT<1,"Cash flow could be a problem.",IF (RatioAT=1,"Cash flow adequate.","Good cash flow."))
Operating ratio (RatioOP)	=(COGS+Expenses)/NetSales	=IF(RatioOP<1,"Desirable!",IF(RatioOP=1,"Marginal.","Undesirable."))
Gross profit margin ratio (RatioGP)	=GrossProfit/NetSales	=IF(AND(RatioGP>0.25,RatioGP<0.4),"Industry Average",IF(RatioGP>1, "ERROR","Uncommonly high profit margin. "))
Asset turnover ratio (RatioAT)	=NetSales/TotalAssets	=IF(AND(RatioAT>.04,RatioAT<.1),"Industry Average","High sales to assets ratio. ")
Total debt to total assets ratio (RatioTDTA)	=TotalLia/TotalAssets	=IF(RatioTDTA<0.75, "Industry Average",IF (AND(RatioTDTA>=.75, RatioTDTA<=1),"Debt ratio too high.", "Debts exceed assets. "))

Current Ratio

This business' current ratio tells business owners, investors, and lenders whether a business has enough assets to cover its liabilities. Assets left over after subtracting liabilities is called working capital. By dividing the assets by the liabilities you will see how many dollars the business has in assets for every dollar of liabilities.

Exercise 20

To set up the current ratio portion of your financial statements worksheet:

1. Open the FINSTATE spreadsheet.
2. Enter the labels **Sander's Woodworks, 1996 Current Ratio Report, Jun-96,** *blank row,* **Current Assets, Current Liabilities,** *blank row,* **Working Assets,** *blank row,* **Ratio,** *blank row,* **Analysis**. The row headings are bold and right aligned. The name in R103 is derived from a reference to the Name range: =NAME. The month and year refers to the MonthYr range: =MONTHYR.
3. Bold and center the titles across the R and S columns.
4. Select and name the range R103:S114 as the CURRENTRATIO range.
5. Format columns S107, S108, and S110 as currency.
6. Enter the range name reference to Total Current Assets and Total Current Liabilities in S107 and S108 respectively: TOTALCURRENTASSETS and TOTALCURRENTLIA.
7. Enter the formula for Working Capital in cell S110: =TOTALCURRENTASSETS-TOTALCURRENTLIA.
8. Enter the formula for the current ratio: =TOTALCURRENTASSETS/ TOTALCURRENTLIA.
9. Select the resulting ratio cell and name it the RATIOCR range. (This step will help support our final summarizing exercise.)
10. Format the spreadsheet as shown in Figure 3-12.
11. Save the spreadsheet.

Exercise 21

The current ratio's value can be interpreted quite easily since a value less than one would indicate more debt than assets (less than a one to one ratio), one would be equal, and over one shows your assets are greater than debts (more than a one to one ratio). Experience in many industries shows that two or more is desirable. To enter an analysis of the Sander's Woodworks current ratio:

1. In cell S114, create a nested IF function that reports on the meaning of the ratio (from Table 3-3): =IF(RatioCR<1,"Debts greater than assets.",IF(RATIOCR=1,"Assets equal to Debts.",IF(RatioCR>=2,"Desirable!","Assets greater than debts.")))
2. Save the spreadsheet.
3. Save the spreadsheet again as the SANDERS spreadsheet.
4. Copy the CurrentRatio range to your Sander's Report document and write a paragraph explaining the meaning of the current ratio's results.
5. Save the spreadsheet.

Compare your work to Figure 3-12 on page 44.

CurrentRatio | =TotalAssets96/TotalLia96

	R	S	T
103	Sander's Woodworks		
104	1996 Current Ratio Report		
105	Jun-96		
106			
107	Current Assets	$ 38,012.00	
108	Current Liabilities	$ 15,256.00	
109			
110	Working Assets	$ 22,756.00	
111			
112	Ratio	2.491609858	
113			
114	Analysis	Desirable!	
115			

Figure 3-12: Sander's Woodworks gets a Desirable! rating with a 9 to 1 current ratio.

Acid-Test Ratio

The Acid-Test ratio is very similar to the Current Ratio except that it answers the question: If the business needed to pay all of its creditors in a short period of time, could it do so? The way to answer this question is to know how much money the business could raise quickly from its assets. Those assets that qualify must be turned into cash quickly and are called Quick Current Assets and are the result of subtracting the Inventory's value from the Total Current Assets value. The Acid-Test ratio divides the quick current assets by the Total Current Liabilities.

Exercise 22

To create a worksheet and report form for the Acid-Test ratio:

1. Open the FINSTATE spreadsheet.
2. To fill in the Acid-Test ratio portion of the spreadsheet, starting in cell W103, enter the labels **Sander's Woodworks, 1996 Acid-Test Ratio Report , Jun-96,** *blank row,* **Current Assets, Inventory,** *blank row,* **Quick Current Assets, Current Liabilities,** *blank row,* **Ratio,** *blank row,* **Analysis.**
3. Enter the Current Assets value (cell X107) with a reference to the TOTALCURRENTASSETS named range in the balance sheet: =TOTALCURRENTASSETS.
4. Enter the Inventory value (cell X108) with a reference to the INVENTORY named range in the balance sheet: =INVENTORY.
5. Enter the formula for Quick Current Assets: =TOTALCURRENTASSETS-INVENTORY.
6. Enter the Acid-Test ratio formula: =X110/TOTALLIA
7. Name the cell with the resulting figure the RATIOAT range.
8. Name the range W103:X115, ACIDRATIO.
9. Format the spreadsheet as shown in Figure 3-13.
10. Save the spreadsheet.

Exercise 23

The analysis for the Acid-Test ratio is similar to the Current Ratio. Values less than 1 show a possible problem with cash flow in that the organization would not be able to cover its debt quickly. A value of one shows a one-to-one ratio between debt and assets, which is satisfactory. A value over one shows a good financial condition for most businesses. To enter the Acid-Test analysis:

1. In cell X115 create a nested IF function (Table 3-3) that reports on the meaning of the ratio: =IF(RATIOAT<1,"Cash flow could be a problem.",IF(RATIOAT=1,"Cash flow adequate.","Good cash flow."))
2. Save the spreadsheet.
3. Save the spreadsheet again as the SANDERS spreadsheet.
4. Copy the AcidRatio range to your Sander's Report document and write a paragraph explaining the meaning of the Acid-Test ratio's results.

AcidRatio | =X110/TotalLia96

	W	X	Y	Z
103	**Sander's Woodworks**			
104	**1996 Acid-Test Ratio Report**			
105	**Jun-96**			
106				
107	**Current Assets**	$ 38,012.00		
108	**Inventory**	$ 17,317.00		
109				
110	**Quick Current Assets**	$ 20,695.00		
111	**Current Liabilities**	$ 18,287.00		
112				
113	**Ratio**	1.356515469		
114				
115	**Analysis**	Business is in good financial condition.		
116				

Figure 3-13: The Acid-Test Ratio for Sander's Woodworks.

Ratios Based on Sales and Other Ratios

A business' sales must cover its cost of operating. The Operating ratio takes the cost of goods and administrative expenses into account when trying to judge the profitability of a business. It is also an example of a ratio based on sales. The Gross Profit margin ratio and the Asset Turnover ratio are also interpreted in relation to Net Sales. The Total Debt to Total Assets ratio compares the total liabilities to the total assets showing the total indebtedness of an organization for each $1 in assets. Follow the directions in Exercise 21 to complete an Operating Ratio portion to the FINSTATE spreadsheet, and then complete your template by including portions for the other ratios.

Exercise 24

To add a business' operating ratio to your spreadsheet:

1. Open the FINSTATE spreadsheet.
2. Enter the labels **Sander's Woodworks, 1996 Operating Ratio Report, Jun-96,** *blank row,* **COGS, Operating Expenses,** *blank row,* **Net Sales,** *blank row,* **Ratio,** *blank row,* **Analysis.**
3. Name the active area the OPERATINGRATIO range.

4. Enter the Operating Ratio formula: =(cost of goods sold + operating expenses)/net sales. (See Table 3-3 on page 42.)

5. Name the resulting cell the RATIOOP range.

6. Enter the analysis IF function formula to reflect the results for the Operating Ratio from Table 3-3: =IF(RATIOOP<1,"Desirable!",IF(RATIOOP=1, "Marginal.","Undesirable."))

7. Format the spreadsheet as shown in Figure 3-14..

8. Save the spreadsheet.

The results should look like Figure 3-14.

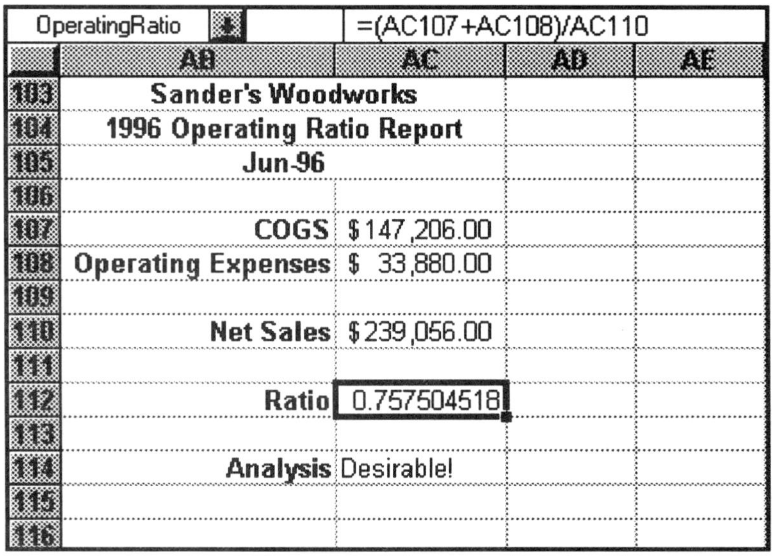

Figure 3-14: The Current Ratio worksheet also shows Sander's Woodworks in desirable shape.

Exercise 25

Given the formulas and analyses for the remaining three ratios (Table 3-3), include sections for them in your financial report document. Name their ranges the GROSSRATIO and RATIOGP, ASSETRATIO and RATIOAT, and TOTALRATIO and RATIOTDTA.

Summary of Ratio Data

As a last step in preparing your financial report, you can summarize the ratios to give the business owner or investor a quick reference to the business' condition.

Exercise 26

To add a report to show a business' ratio results, recreate Table 3-4 below. Use the named ranges to complete the report. (***Hint**: How can you use named ranges to fill in the analysis portion of the table?*) Add this table to your financial document and write a paragraph based on its contents. Save your template for use in the coming application exercises.

Table 3-4

Sander's Woodworks Financial Ratio Report TODAY()		
Ratio	**Value**	**Analysis**
Current	9.37	Desirable!
Acid-test	2.55	Business is in good financial condition.
Operating	0.76	Desirable!
Gross Profit Margin	0.38	Industry Average
Asset Turnover	6.29	Industry Average
Total Debt to Total Assets	0.40	Lower than average.

Exercise 27

Prepare the FINSTATE spreadsheet as a template. Select the whole spreadsheet and protect it. Then find and unlock each data entry cell. Save it as the FINTEMPL spreadsheet. Remember to use the Save As command each time you start a new financial statement spreadsheet.

Application Exercises

Use the sample spreadsheet and worksheets you made in this chapter to complete the following exercises. You can quickly go to your worksheets by using the Go To feature and the range names.

1. Open the FINTEMPL spreadsheet and save a version for each of the following balance sheet examples. Complete each balance sheet:

a. FAWCETT

Fawcett's Plumbing Supplies	
=TODAY()	
Assets	
Current Assets:	
Cash	$ 1,724.00
Office supplies	$ 173.00
Accounts receivable	$ 9,374.00
Plant Assets:	
Equipment	$ 12,187.00
Liabilities	
Current Liabilities:	
Accounts payable	$ 2,174.00
Wages payable	$ 674.00
Property and taxes payable	$ 250.00
Equity	
D. Fawcett Capital	$ 20,610.00

b. ROOTER

Rooter Company Balance Sheet	
=TODAY()	
Assets	
Current Assets:	
Cash	$ 2,350.00
Supplies	$ 175.00
Accounts receivable	$ 8,956.00
Plant Assets:	
Equipment	$ 11,375.00
Liabilities	
Current Liabilities:	
Accounts payable	$ 1,940.00
Wages payable	$ 855.00
Rent payable	$ 775.00
Equity	
Wilson Rooter Capital	$ 20,061.00

c. MIDLAND

Midland Products, Inc. Balance Sheet	
=TODAY()	
Assets	
Current Assets:	
Checking	$ 4,500.00
Money-market account	$ 18,000.00
Accounts receivable	$ 45,000.00
Inventory	$112,500.00
Plant Assets:	
Plant site	$ 72,000.00
Building and equipment	$186,000.00
Delivery trucks	$ 85,000.00
Liabilities	
Current Liabilities:	
Accounts payable	$ 54,000.00
Notes payable	$ 18,000.00
Long-term Liabilities:	
Mortgage	$ 54,000.00
Notes payable	$ 36,000.00
Equity	
Midland Products Capital	$258,000.00

d. ROYRUSS

Roy Russell's Security Service	
=TODAY()	
Assets	
Current Assets:	
Cash	$ 8,000.00
Accounts receivable	$ 4,860.00
Inventory	$19,823.00
Equipment	$ 8,925.00
Liabilities	
Current Liabilities:	
Accounts payable	$11,281.00
Wages payable	$11,185.00
Owner's Equity	$19,142.00

2. For the spreadsheetS you made in Application Exercise 1, prepare a vertical analysis for the three companies as well as for Scott Pharmaceutical Co., Inc. and Lush Lawns, Inc. below. ***Note; depreciation is a negative number.***

a. SCOTT

Scott Pharmaceutical Co., Inc.	
=TODAY()	
Assets	
Current Assets:	
Cash	$ 5,800.00
Accounts receivable	$ 11,200.00
Supplies	$ 500.00
Inventory	$ 19,000.00
Fixed Assets:	
Equipment	$ 44,000.00
Less depreciation	$ (9,000.00)
Building	$ 400,000.00
Less depreciation	$ (35,000.00)
Land	$ 200,000.00
Liabilities	
Current Liabilities:	
Notes payable	$ 8,000.00
Accounts payable	$ 10,000.00
Long-term Liabilities:	
Mortgage note payable	$ 84,000.00
Equity	
Common stock	$ 259,000.00
Retained earnings	$ 239,500.00

b. LUSH

Lush Lawns, Inc.		
=TODAY()		
Assets		
Current Assets:		
Cash	$	50,000.00
Accounts receivable	$	24,300.00
Notes receivable	$	10,000.00
Inventory	$	33,000.00
Fixed Assets:		
Land (at cost)	$	31,500.00
Building (at cost)	$	45,000.00
Machinery, equipment (at cost)	$	34,000.00
Less depreciation	$	(30,000.00)
Liabilities		
Current Liabilities:		
Notes payable	$	17,500.00
Accounts payable	$	20,400.00
Accrued expenses	$	5,300.00
Accrued taxes	$	7,200.00
Long-term Liabilities:		
Mortgage payable	$	49,000.00
Equity		
Capital stock	$	58,000.00
Retained earnings	$	40,400.00

3. Using the spreadsheets created thus far for the Application Exercises, prepare comparative balance sheets for the above companies given the following figures from the same time period one year earlier: =TODAY()-365.

a.

Fawcett's Plumbing Supplies Balance Sheet	
=TODAY()-365	
Assets	
Current Assets:	
Cash	$ 1,396.00
Office supplies	$ 273.00
Accounts receivable	$ 8,377.00
Plant Assets:	
Equipment	$ 11,578.00
Liabilities	
Current Liabilities:	
Accounts payable	$ 2,256.00
Wages payable	$ 607.00
Property and taxes payable	$ 260.00
Equity	
D. Fawcett Capital	$ 18,501.00

b.

Rooter Company	
=TODAY()-365	
Assets	
Current Assets:	
Cash	$ 3,750.00
Supplies	$ 178.00
Accounts receivable	$ 6,556.00
Plant Assets:	
Equipment	$ 9,354.00
Liabilities	
Current Liabilities:	
Accounts payable	$ 1,250.00
Wages payable	$ 755.00
Rent payable	$ 675.00
Equity	
Wilson Rooter Capital	$ 17,178.00

c.

Midland Products, Inc. =TODAY()-365	
Assets	
Current Assets:	
Checking	$ 3,300.00
Money-market account	$ 16,000.00
Accounts receivable	$ 35,000.00
Inventory	$124,500.00
Plant Assets:	
Plant site	$ 65,000.00
Building and equipment	$186,000.00
Delivery trucks	$ 105,000.00
Liabilities	
Current Liabilities:	
Accounts payable	$ 57,000.00
Notes payable	$ 22,000.00
Long-term Liabilities:	
Mortgage	$ 54,000.00
Notes payable	$ 48,000.00
Equity	
Midland Products Capital	$401,800.00

d.

Roy Russell's Security Service =TODAY()-365	
Assets	
Current Assets:	
Cash	$ 7,000.00
Accounts receivable	$ 995.00
Inventory	$22,571.00
Equipment	$ 8,655.00
Liabilities	
Current Liabilities:	
Accounts payable	$14,281.00
Wages payable	$ 9,185.00
Owner's Equity	$15,755.00

e.

Scott Pharmaceutical Co., Inc. =TODAY()-365	
Assets	
Current Assets:	
Cash	$ 7,800.00
Accounts receivable	$ 14,300.00
Supplies	$ 400.00
Inventory	$ 25,000.00
Fixed Assets:	
Equipment	$ 48,000.00
Less depreciation	$ (7,000.00)
Building	$ 400,000.00
Less depreciation	$ (35,000.00)
Land	$ 200,000.00
Liabilities	
Current Liabilities:	
Notes payable	$ 13,000.00
Accounts payable	$ 15,000.00
Long-term Liabilities:	
Mortgage note payable	$ 84,000.00
Equity	
Common stock	$ 266,500.00
Retained earnings	$ 108,000.00

f.

Lush Lawns, Inc. =TODAY()-365	
Assets	
Current Assets:	
Cash	$ 57,000.00
Accounts receivable	$ 20,500.00
Notes receivable	$ 18,000.00
Inventory	$ 38,000.00
Fixed Assets:	
Land (at cost)	$ 31,500.00
Building (at cost)	$ 45,000.00
Machinery, equipment (at cost)	$ 29,000.00
Less depreciation	$ (27,000.00)
Liabilities	
Current Liabilities:	
Notes payable	$ 15,500.00
Accounts payable	$ 21,800.00
Accrued expenses	$ 7,400.00
Accrued taxes	$ 7,200.00
Long-term Liabilities:	
Mortgage payable	$ 49,000.00
Equity	
Capital stock	$ 63,000.00
Retained earnings	$ 57,400.00

4. Using the spreadsheets created thus far in the Application Exercises, prepare income statements for for each company given the following data.

a.

Fawcett's Plumbing Supplies Income Statement for period Jan 1 1996-Mar 31 1996	
Gross sales	$238,923.00
Sales returns and allowances	$ 13,815.00
Beginning inventory, (Jan-1, 1995)	$ 25,814.00
Purchases	$109,838.00
Ending inventory, (Mar-31, 1995)	$ 23,423.00
Salary	$ 42,523.00
Insurance	$ 8,640.00
Utilities	$ 860.00
Rent	$ 8,640.00
Depreciation	$ 1,920.00
Miscellaneous	$ 3,420.00

b.

Rooter Company Income Statement for period Jan 1 1996-Mar 31 1996	
Gross sales	$148,645.00
Sales returns and allowances	$ 8,892.00
Beginning inventory, (mm-1)	$ 12,100.00
Purchases	$ 47,800.00
Ending inventory, (mm-31)	$ 11,950.00
Salary	$ 25,500.00
Insurance	$ 2,100.00
Utilities	$ 1,445.00
Rent	$ 4,500.00
Depreciation	$ 1,240.00
Miscellaneous	$ 750.00
Fees	$ 225.00

c.

Midland Products, Inc Income Statement for period Jan 1 1996-Mar 31 1996	
Gross sales	$291,707.00
Sales returns and allowances	$ 5,895.00
Beginning inventory, (mm-1)	$ 35,892.00
Purchases	$157,213.00
Ending inventory, (mm-1)	$ 32,516.00
Salary	$ 42,000.00
Insurance	$ 3,800.00
Utilities	$ 1,986.00
Rent	$ 3,600.00
Depreciation	$ 4,000.00

d.

Roy Russell's Security Service Income Statement for period Jan 1 1996-Mar 31 1996	
Gross sales	$239,873.00
Sales returns and allowances	$ 12,815.00
Beginning inventory, (mm-1)	$ 27,814.00
Purchases	$123,213.00
Ending inventory, (mm-31)	$ 24,482.00
Salary	$ 44,772.00
Insurance	$ 3,006.00
Utilities	$ 1,597.00
Rent	$ 3,600.00
Depreciation	$ 4,100.00

e.

Scott Pharmaceutical Co., Inc. Income Statement for period Jan 1 1996-Mar 31 1996	
Gross sales	$219,827.00
Sales returns and allowances	$ 8,512.00
Beginning inventory, (mm-1)	$ 42,816.00
Purchases	$ 97,523.00
Ending inventory, (mm-31)	$ 43,182.00
Salary	$ 28,940.00
Insurance	$ 800.00
Utilities	$ 1,700.00
Rent	$ 3,600.00
Depreciation	$ 2,000.00

f.

Lush Lawns, Inc. Income Statement for period Jan 1 1996-Mar 31 1996	
Gross sales	$ 28,000.00
Sales returns and allowances	$ 500.00
Beginning inventory, (mm-1)	$ 6,000.00
Purchases	$ 20,500.00
Ending inventory, (mm-31)	$ 7,500.00
Administration	$ 5,000.00
Occupancy	$ 750.00
Sales	$ 1,600.00
Miscellaneous	$ 150.00

5. Using the Application Exercise spreadsheets, prepare comparative income statements with horizontal analyses for all the above given the new data below.

a.

Fawcett's Plumbing Supplies Income Statement for period Jan 1 1995-Mar 31 1995	
Gross sales	$ 225,863.00
Sales returns and allowances	$ 10,586.00
Beginning inventory (mm-1)	$ 22,925.00
Purchases	$ 100,633.00
Ending inventory (mm-31)	$ 21,470.00
Salary	$ 44,663.00
Insurance	$ 2,566.00
Utilities	$ 996.00
Rent	$ 8,640.00
Depreciation	$ 997.00
Miscellaneous	$ 2,544.00

b.

Rooter Company Income Statement for period Jan 1 1995-Mar 31 1995	
Gross sales	$ 197,233.00
Sales returns and allowances	$ 9,952.00
Beginning inventory (mm-1)	$ 11,099.00
Purchases	$ 44,300.00
Ending inventory (mm-31)	$ 9,765.00
Salary	$ 31,920.00
Insurance	$ 2,100.00
Utilities	$ 1,862.00
Rent	$ 4,500.00
Depreciation	$ 9,964.00
Miscellaneous	$ 825.00
Fees	$ 180.00

c.

Midland Products, Inc Income Statement for period Jan 1 1995-Mar 31 1995	
Gross sales	$ 275,873.00
Sales returns and allowances	$ 6,821.00
Beginning inventory (mm-1)	$ 32,587.00
Purchases	$ 146,999.00
Ending inventory (mm-1)	$ 30,013.00
Salary	$ 40,000.00
Insurance	$ 3,800.00
Utilities	$ 2,097.00
Rent	$ 3,300.00
Depreciation	$ 4,500.00

d.

Roy Russell's Security Service Income Statement for period Jan 1 1995-Mar 31 1995	
Gross sales	$ 236,941.00
Sales returns and allowances	$ 13,895.00
Beginning inventory (mm-1)	$ 25,887.00
Purchases	$ 112,604.00
Ending inventory (mm-31)	$ 23,838.00
Salary	$ 42,640.00
Insurance	$ 2,863.00
Utilities	$ 1,521.00
Rent	$ 3,600.00
Depreciation	$ 3,400.00

e.

Scott Pharmaceutical Co., Inc. Income Statement for period Jan 1 1995-Mar 31 1995	
Gross sales	$ 205,852.00
Sales returns and allowances	$ 7,983.00
Beginning inventory (mm-1)	$ 40,512.00
Purchases	$ 94,812.00
Ending inventory (mm-31)	$ 42,521.00
Salary	$ 27,000.00
Insurance	$ 750.00
Utilities	$ 1,580.00
Rent	$ 3,000.00
Depreciation	$ 2,400.00

f.

Lush Lawns, Inc. Income Statement for period Jan 1 1995-Mar 31 1995	
Gross sales	$ 26,000.00
Sales returns and allowances	$ 670.00
Beginning inventory (mm-1)	$ 7,000.00
Purchases	$ 16,500.00
Ending inventory (mm-31)	$ 6,000.00
Administration	$ 4,500.00
Occupancy	$ 500.00
Sales	$ 1,000.00
Miscellaneous	$ 250.00

6. Using the Appliation Exercise spreadsheets, prepare a financial ratio summary document as in Table 3-4 for all the above (application exercises 1 to 5) given the available data.

II. Lending and Borrowing

4. Simple Interest

At first glance, the simple interest formula of Principal x Rate x Time seems...well, simple. However, it becomes complicated when we talk about loans for less than one year and time based on the 360- versus the 365-day per year methods or when comparing various loan options. The spreadsheet can help you keep the "simple" in simple interest.

Objectives

By working through and completing the exercises below, you will be able to:

- Prepare a simple interest loan spreadsheet.
- Prepare a spreadsheet that compares different loan offers.
- Use the 360-day method to compute interest for loans whose time is in days.
- Use the 365-day method to compute interest for loans whose time is in days.
- Prepare a simple interest loan template.

Simple Interest on Loans for Years or for Months

Simple interest loans are usually short-term loans of one year or less but they can be for longer periods as well. The interest is added to the principal to find the loan's maturity value, which is the amount the borrower pays at the end of the time period.

A Basic Simple Interest Loan of Less than One Year

To compute a simple interest loan of less than one year but measured in months, you will only need to use three formulas in your spreadsheet. Time in months must be divided by 12 to find the percent of a year (Time in Years) value. The interest is the multiplication of principal, rate, and time. The maturity value is the sum of principal and interest.

Exercise 1

Consider a simple loan to George Jones for the following spreadsheet, which you can use as a template for future simple interest loans. George's father-in-law will lend George $900 to purchase parts for his antique car. They agree on 16% interest for a period of five months.

To find out how much George will have to pay his father-in-law at the end of the five months:

1. Create a new spreadsheet and save it as the LENDING spreadsheet.
2. Beginning in cell A1 and moving down the column, enter **Simple Interest Loan for, George Jones, Principal, Rate, Time in Months**, *blank row*, **Time in Years**, *blank row*, **Interest**, *blank row*, **Maturity Value.**
3. Bold and center these titles across columns A and B.
4. Bold and right align the labels.
5. Fill in the amounts from the problem statement above for Principal, Rate, and Time in Months.
6. Format the amounts appropriately.
7. Enter the formula for Time in Years: =Time in months/12, using the reference to the Time in months value's cell; in our example it is =B6/12
8. Enter the formula for Interest: =Principal*Rate*Time; in our example it is =B4*B5*B8.

9. Enter the formula for Maturity Value: =Principal + Interest; in our example it is =B4+B10.
10. Save the spreadsheet.

Your spreadsheet should look like Figure 4-1 below.

A1 | Simple Interest Loan for

	A	B	C	D
1	Simple Interest Loan for			
2	George Jones			
3				
4	Principal	$ 900.00		
5	Rate	16.0%		
6	Time in Months	5		
7				
8	Time in Years	0.416667		
9				
10	Interest	$ 60.00		
11				
12	Maturity Value	$ 960.00		
13				

Figure 4-1: George Jones will need to pay his father-in-law the principal of $900 and $60 interest in five months.

Exercise 2

Use the information for George Jones to make a simple loan document his father-in-law could use if he wanted to have a written agreement. Create it in a word processing program. For example it might start with, "I _________________ agree to pay the principal of _____ ..."

Finding the Best Deal

As is true of most of life's situations, when we want to borrow money there is usually more than one possible source with different terms. In such a case you will want to compute all the offered deals and find the best overall deal. The meaning of "best deal" will depend on your priorities regarding amount needed, at what rate of interest, and with how much time to pay.

Exercise 3

Use your LENDING simple interest template to find the best deal for George. George wants as much time to pay as he can get but it must not be less than three months. He also does not want to pay over $50 in interest. Using the information in Table 4-1, find George's best offer.

Table 4-1

Principal	Rate	Time
$ 900.00	16.0%	5
$ 900.00	14.0%	9
$ 900.00	18.0%	4
$ 900.00	10.5%	10
$ 900.00	15.0%	6
$ 900.00	9.9%	3

1. Open the LENDING spreadsheet.

2. Copy the Simple Interest portion of the spreadsheet (A1:B12) to AA1.
3. Fill in the amounts from Table 4-1 for Principal, Rate, and Time in months across the columns.
4. Copy the formulas for Time in Years, Interest, and Maturity Value to the right beneath all the amounts.
5. Save the spreadsheet.

Your spreadsheet should look like Figure 4-2:

BestDeal | Simple Interest Best Deal

	AA	AB	AC	AD	AE	AF	AG
1	Simple Interest Best Deal						
2	George Jones						
3							
4	Principal	$ 900.00	$ 900.00	$ 900.00	$ 900.00	$ 900.00	$ 900.00
5	Rate	16.0%	14.0%	18.0%	10.5%	15.0%	9.9%
6	Time in Months	5	9	4	10	6	3
7							
8	Time in Years	0.416666667	0.75	0.333333333	0.833333333	0.5	0.25
9							
10	Interest	$ 60.00	$ 94.50	$ 54.00	$ 78.75	$ 67.50	$ 22.28
11							
12	Maturity Value	$ 960.00	$ 994.50	$ 954.00	$ 978.75	$ 967.50	$ 922.28
13							

Figure 4-2: What is George's best deal on a simple interest loan?

Simple Interest on Loans with Time in Days.

When time is measured in days, the simple can seem to escape the Simple Interest loan calculation. With the era of the spreadsheet and computers, the pencil and paper formulas of the past have given way to two possible ways to compute interest. They are based on a 360-day (30 days per month) year and a 365-day year. The worksheet you will create must also meet your needs for a report.

Loans Based on a 360-day Year

When calculating the difference between two dates, a spreadsheet uses exact time (365 or 366 days). To have it use a 360-day year, you must use the DAYS360 function. In the Number of Days column, you must use the DAYS360 function referencing the beginning and due dates. You will then use this amount divided by the number of days in the year (360) to find the Time in Years:

Number of Days = DAYS360(Beginning Date,Ending Date)

Time in Years = Number of Days/Method's Number of Days in Year

Exercise 4

To create a worksheet that calculates a loan using Ordinary Time and Ordinary Interest, use an example for a $730 loan made on March 1 and due December 1 at a 12% interest rate:

1. Open the LENDING spreadsheet and start in a cell well away from the above work or start a new worksheet (in multiple worksheet programs like Microsoft Excel, Lotus 1-2-3, or Borland's Quattro Pro).
2. Enter the title of this portion of the spreadsheet: **Simple Interest Loan.**
3. Skip a row and enter the labels: **Principal, Interest Rate,** *blank row,* **Beginning Date,** *blank row,* **Due Date,** *blank row,* **Method,** *blank row,* **Number of Days, Time in Years,** *blank row,* **Interest,** *blank row,* **Maturity Value.**
4. Enter the data from the problem description given above and 360 for the Method.

5. Enter the function for finding the number of days. In our example it is =DAYS360(B7,B9).
6. Enter the formula for the Time in Years. In our example it is =B13/B11.
7. Enter the formula for the Interest. In our example it is =B4*B5*B14.
8. Enter the formula for the Maturity Value. In our example it is =B4+B16.
9. Save the spreadsheet.

Compare your results with Figure 4-3.

A1 | Simple Interest Loan

	A	B	C	D
1	**Simple Interest Loan**			
2				
3				
4	**Principal**	$ 730.00		
5	**Interest Rate**	$ 0.12		
6				
7	**Beginning Date**	3/1/96		
8				
9	**Due Date**	12/1/96		
10				
11	**Method**	**360**		
12				
13	**Number of Days**	270		
14	**Time in Years**	75.0%		
15				
16	**Interest**	$ 65.70		
17				
18	**Maturity Value**	$ 795.70		
19				

Figure 4-3: A simple interest loan using 360-day method to calculate interest.

Loans Based on a 365-day Year

Exercise 5

The second method in use today for computing interest on a simple interest loan is the 365-days-per-year method. To set up its template:

1. Copy all the cells in the 360-day method example one cell to the right
2. Substitute 365 for the Method entry.
3. Substitue a formula to find the exact number of days: Ending Date - Beginning Date, which in our example is =C9-C7.
4. Save the spreadsheet.

Consult Figure 4-4 to confirm the results.

C13 =C9-C7

	A	B	C	D
1	**Simple Interest Loan**			
2				
3				
4	**Principal**	$ 730.00	$ 730.00	
5	**Interest Rate**	$ 0.12	$ 0.12	
6				
7	**Beginning Date**	3/1/96	3/1/96	
8				
9	**Due Date**	12/1/96	12/1/96	
10				
11	**Method**	**360**	**365**	
12				
13	**Number of Days**	270	275	
14	**Time in Years**	75.0%	75.3%	
15				
16	**Interest**	$ 65.70	$ 66.00	
17				
18	**Maturity Value**	$ 795.70	$ 796.00	
19				

Figure 4-4: The 365-day method for computing simple interest yields more interest than the 360-day method.

Exercise 6

Complete the process and make the LENDING spreadsheet into a template. Unlock the data entry cells and then protect the entire spreadsheet. Save it as the LENDTEMP spreadsheet.

Application Exercises

Use your simple interest spreadsheet template to answer the following:

1. Find the interest and maturity value for the following loans using simple interest. Use your simple interest spreadsheet and then prepare and print a loan agreement on each. Use your name as the borrower.

    a. $600 at 12% for 1 year.

    b. $700 at 10% for 6 months.

    c. $600 at 11% for 7 months.

    d. $450 at 9.5% for 8 months.

    e. $870 at 10% for 10 months.

    f. $500 at 9% for 1 year.

    g. $500 at 9% for 6 months.

    h. $780 at 12% for 10 months.

    i. $900 at 10.5% for 8 months.

    j. $840 at 14% for 5 months.

2. Find the interest paid on a loan of $24,000 for one year at a simple interest rate of 11% per year. What is the maturity value?

3. Find the interest paid on a loan of $800 at 8.5% annual simple interest for two years. What is the maturity value?

4. How much interest will have to be paid on a loan of $7,980 for two years at a simple interest rate of 6.2% per year? What is the maturity value?

5. Find the total amount of money that the borrower will pay back on a loan of $1,400 at 12.5% annual simple interest for 3 years. How much of that is interest?

6. Find the maturity value of a loan of $2,800 after three years. The loan carries a simple interest rate of 7.5% per year.

7. Raul Fletes needs money to buy lawn equipment. If he borrowed $500 for seven months, what rate should he bargain for so as to not pay more than $40 in interest? (Experiment with different rates.)

8. Rob Thweat needs money for college. He must borrow $6,000 at 12% simple interest per year. If he wants to keep his interest between $300 and $360, what are the longest and shortest target durations of the loan? (Experiment with different durations.)

9. Linda Davis agreed to lend money to Alex Luciano at a special interest rate of 9% per year, on the condition that he borrow enough that he would pay her $500 in interest over a two-year period. What is the mininum amount Alex could borrow? (Experiment with different amounts.)

10. Keaton Smith wants to borrow $25,000 to purchase stock for his baseball card shop. He expects he can repay the simple interest loan after three years. He wants to keep the interest down to $6,000. What interest rates should he agree to? (Experiment with different rates.)

11. A loan for $3,000 with simple annual interest rate of 15% was made on June 15 and was due on August 15. What is the maturity value using the 360-day method? With the 365-method?

12. What is the duration of a loan of $3,000 if the loan required interest of $416.25 and was at a rate of 18.5% annual simple interest. (Create a formula for fionding the duration from the interest rate and prinipal or experiment with different durations.)

13. Find the rate of simple interest on a $600 loan with total interest of $40.50 if the loan is paid in 6 months. What is the maturity value using the 360-day method? With the 365-method?

14. Find the interest on a loan of $25,000 at 11.75% annually for 21 days. Which gives the lender more profit, the 360- or 365-day method? Use your new formula or experiment. Prepare a short report on each.

15. Find the interest paid on a loan of $1,600 for 90 days at a simple interest rate of 16% annually. Which gives the borrower the best deal, the 360- or the 365-day method? Prepare a short report on each.

16. Find the interest on a loan of $1,510 at 12.75% annual interest for 27 days. What is the maturity value using the 360-day method? With the 365-day method?.

17. Find the interest on a government loan of $4,300 at 11.75% annual interest for 32 days. Use the 365-day method.

18. Terrific Toys needs to borrow $140,000 to finance an inventory for Christmas sales. The loan will be for 85 days at 11.5%. How much more would the interest be if the loan were computed by the 365-day method, rather than the 360-day method? Print a report showing each.

19. Dan Dash needs $168 for the repair of the car that gets him to school. His father agrees to lend him the money, but to discourage Dan from borrowing, his father adds $1.54 a month until the debt is paid. What interest rate is Dan's father charging? Use the 360-day method. (***Hint:*** *What rate would give $1.54 per month and how many days in a month?*)

20. A loan shark made a loan of $960 and was repaid $1,000 the following month. What rate of interest was the loan shark charging? Use the 360-day method.

5. Bank Discounts

Banks often provide a discount note rather than a simple interest note. With discount notes the borrower pays the interest on a loan up front, which makes the interest the discount. In this way the bank makes sure it gets its profit and the user needs to pay back the remaining amount. The borrower gets to use only a portion of the face value of the note and this portion is called the proceeds.

Simple interest notes (promissory notes) and discount notes are legally binding documents. Even verbal agreements are binding but written agreements can be transferred through a process called rediscounting. This is usually done through interested investors. The result is that the lender gets less than the maturity value and the borrower pays a new creditor, the purchaser of the note. The borrower does not pay more or less than originally agreed upon. The buyers of notes make a profit by purchasing the notes at a higher discount rate than the original note's interest rate. In other words, the original lender receives less interest than originally planned.

Objectives:

By working through and completing the exercises below, you will be able to:

- Prepare a template to compute a simple discount note.
- Use the spreadsheet to calculate the face value of a simple discount note.

Simple Discount Notes

For a simple interest note (loan) the face value of the note is the principal (the amount the borrower gets to use). By adding the interest you are able to find the maturity value (the amount the borrower must repay). For a simple discount note the face value is the maturity value and the interest is the discount. To find the proceeds (the amount the borrower gets to use) you must subtract the discount from the face value.

The labels for a simple discount worksheet resemble a simple interest worksheet. The data includes the face value (maturity value), the discount rate, and the time (we will use the 365-day method). The formulas compute the discount and the proceeds. The formula for the proceeds differs from the simple interest maturity value in that you must subtract the discount from the face value (in simple interest you add the interest to the principal).

Exercise 1

Set up the title and labels for your simple discount worksheet:

1. Open a new spreadsheet and save it as the DISCOUNT spreadsheet.
2. Enter the title **Calculation of Proceeds from Face Value.**
3. Skip two rows and enter the column labels **Name, Face Value, Rate, Begin, Due, No Days, Time in Years, Discount, Proceeds**.
4. In the cell above the Rate column, enter the label: **Time** and center it across the columns from the Rate column to the Time in Years column.
5. Save the spreadsheet.

Exercise 2

Enter the data for the simple discount worksheet using the following example:

Bill Yurtak wants to borrow money to purchase equipment. The note has a face value of $3,000. The discount rate is 10%. The note is dated March 31 and is due on August 30.

Save the spreadsheet.

Exercise 3

To enter the formulas for the bank discount worksheet:

1. Compute time's number of days. In our example this is =E5-D5.
2. Enter the formula for computing the Time in Years using a 365 day year. In our example this is =F5/365.
3. Use the discount formula (same as the interest formula) to compute the discount. In our example this is =B5*C5*G5.
4. Subtract the discount from the face value to find the proceeds. In our example this is =B5-H5.
5. Format the spreadsheet as shown in Figure 5-1 below and print it making sure to center it on the page.
6. Select all the cells used in the spreadsheet plus one extra row and name it the PROCEEDS range.
7. Save the file.

Your spreadsheet should look like Figure 5-1 below.

A1 | Calculation of Proceeds from Face Value

	A	B	C	D	E	F	G	H	I	J
1				Calculation of Proceeds from Face Value						
2										
3					Time					
4	Name	Face Value	Rate	Begin	Due	No Days	Time in Years	Discount	Proceeds	
5	Value Loan	$ 3,000.00	10%	3/31/96	8/30/96	152	0.416438	$ 124.93	$2,875.07	
6										
7										
8										

Figure 5-1: The bank discount gives borrowers less money to use than what they could have with a simple interest loan.

Calculating the Face Value

Often potential borrowers face the question of how much to borrow. When dealing with discount notes, they must find out the face value based on the amount of money they need to use. The formula to use for this divides the proceeds by one minus the rate times the time; that is:

$$Face\ Value = Proceeds/(1-(Rate * Time))$$

Then, to keep the record straight and for your information, the formula for finding the discount amount is the face value minus the proceeds:

$$Discount = Face\ Value - Proceeds$$

Exercise 4

Use the maturity value spreadsheet to answer the following problem.

A borrower needs $290.80 on a 3-month loan. How much must she borrow if she borrows today at a 12% discount rate?

Discounting a Simple Interest Note

Payees of simple interest notes will sometimes sell them to an investor for cash. The investor will pay proceeds based on the maturity value using its rate and the time until the due date. The interest the payee makes is the difference between the discounted proceeds (what the investor pays) and the principal of the note.

Exercise 5

Consider the following problem and create a Discount spreadsheet to solve it.

Use the simple interest loan computed in Chapter 4 for a 365-day method loan. What would the proceeds be if sold at 14% discount rate on May 11.

1. Open the LENDING spreadsheet and copy the Simple Interest Loan range of cells to a section in the DISCOUNT spreadsheet.
2. Enter the title in the same row and next to the existing title. Enter **Discounting**.
3. Enter the new row labels next to the 365 method entries. Enter **Face Value, Discount Rate**, *blank row*, **Discount Date**, *blank row*, **Due Date**, *blank row*, **Method**, blank *row*, **Number of days, Time in Years**, *blank row*, **Discount**, *blank row*, **Proceeds.**
4. Enter the reference to the face value equal to the note's 365-day maturity value. That should be $796.00 (in our example it is =X43).
5. Fill in the discount rate and discount due date.
6. Copy the rest of the values and formulas by copying the 365-day method's cells from Due Date to Maturity Date.
7. Edit the Proceeds formula to subtract the discount from the Face value. In our example it is =Z29-Z41.
8. Save the spreadsheet.

The lender receives $733.72 for the note and made just $3.72 (the difference between the principal and the proceeds). Your spreadsheet should look like Figure 5-2.

Y26 | Discounting

	V	W	X	Y	Z	AA
26	Simple Interest Loan			Discounting		
27						
28						
29	Principal	$ 730.00	$ 730.00	Face Value	$ 796.00	
30	Interest Rate	12%	12%	Discount Rate	14%	
31						
32	Beginning Date	3/1/96	3/1/96	Discount Date	5/11/96	
33						
34	Due Date	12/1/96	12/1/96	Due Date	12/1/96	
35						
36	Method	360	365	Method	365	
37						
38	Number of Days	270	275	Number of days	204	
39	Time in Years	0.75	0.75	Time in Years	0.56	
40						
41	Interest	$ 65.70	$ 66.00	Discount	$ 62.28	
42						
43	Maturity Value	$ 795.70	$ 796.00	Proceeds	$ 733.72	
44						

Figure 5-2: To rediscount a simple interest note, you must base the discount on the original note's maturity value, which, as opposed to a discounted note, is the principal plus the interest. The proceeds become the maturity value minus the discount.

Exercise 6

Make this spreadsheet into a template file.

Application Exercises

Use your discounting spreadsheets to find the answers to the following problems. Print each report.

1. The owner of the Health Hut decided to install new equipment to sell fresh fruit juices. On August 15 she borrowed $1,600 discounted at 10%. If she borrows for 6 months, what will her proceeds be?

2. What is the face value of a note discounted at 12% for 90 days if the borrower receives $814.80 as proceeds?

3. The proceeds of a 5-month note dated June 15 were $3,330. If the discount rate was 18%, what was the face value of the note?

4. Fancy Furniture Co. owed $876.20 for merchandise plus $10.30 for insurance. The bill was due on May 6 and the owner found she was unable to pay it. To keep a good credit rating, she decided to borrow the necessary amount from a bank that charged a 9% discount. The due date for the loan was July 5. How much did she need to borrow to pay the debt? How much must she repay the bank?

5. On October 15, Terrific Toys borrowed $140,000 to finance an inventory for Christmas sales. The loan was to run for 85 days at 11.5%. If the note were sold after 50 days at a 15% discount rate, what were the proceeds of the rediscounted note? What did Terrific Toys pay? How much did the investor make? Use the 365-day method.

6. First Home Furniture was the payee of a $2,700, 60-day, 10% note dated September 12. On October 22, the note was rediscounted at Madison Collections at 12%. What was the amount originally loaned? What was the amount received when the note was sold? How much did First Home Furniture make on the transaction?

7. Bonus Rooms, a remodeling contractor, was the holder of a $4,000, 108-day note dated June 21, with simple interest at 7.5%. On September 7, Bonus Rooms discounted the note at 12%. How much did the investor pay Bonus Rooms for the note? How much interest did Bonus Rooms earn on the note? How much more would Bonus Rooms have made by keeping the note until maturity?

8. Fresh Flowers Co. had a failure in its heating system and the stock in the hothouse was destroyed. Until the insurance claim was settled they were short of cash and discounted a customer's note. The note was for $7,200 with interest at 6% dated 1/5/90 for 120 days. It was discounted on 3/21/90 at 11%. What were the proceeds.

9. MAK, Inc., accepted an interest-bearing note for $10,000 with 9% annual interest. The note was made on April 10 and was due December 6. MAK needed cash and took the note to an investor, which offered to buy the note at a discount rate of 12.5%. The transaction was made on July 7. How much cash did MAK receive for the note?

10. Brown Trailer Sales made a note of $10,000 with Robert Brown, at 9%. The note was made on August 12 and due on November 10. Brown Trailer Sales needed cash so they took the note to an investor on September 5. The investor accepted the note with a 13% discount rate. How much did Brown Trailer Sales receive from the investor? What was their net interest?

11. City Fruits and Vegetables was the holder of a $1,200, 120-day note discounted at 9%. Thirty days before the note was due, City Fruits and Vegetables rediscounted the note to Capital Mortgage and Trust at 15%. How much did City Fruits and Vegetables receive for the note? How much did it actually make on the transaction?

6. Partial Payments and Installments

Simple interest loans and simple bank discounts are often paid off in portions. When this happens the borrower pays less interest and the lender needs to calculate the interest to date.

Objectives

By working through and completing the exercises below, you will be able to:

- Explain the United States rule.
- Prepare a template to compute the results of partial payments to simple interest notes.
- Make a template to show the results of monthly installment payments.
- Make a template to show the results of revolving credit accounts.
- Prepare a template to show the payment plan for a loan.
- Prepare a template to compute the return on an installment loan paid before its due date and using the Rule of 78.

The United States Rule

The United States rule permeates partial payments from paying simple interest loans off early to revolving charges and installment plans. When a partial payment is made on a loan, it follows four steps:

1. Compute interest on the principal from the first day until the date of the first partial payment.
2. Pay the interest first and then deduct the remaining payment from the principal.
3. On the next payment, calculate interest from the last payment to the present, pay the interest, and deduct the remainder from the remaining balance due.
4. Continue until the maturity date and then compute interest due since the last partial payment and add this interest to the unpaid principal to find the maturity value.

Partial Payment on Simple Interest Notes

The labels, numbers, and formulas for making a partial payment spreadsheet for simple interest loans are already familiar to you. It is simply a matter of keeping track of each successive payment and having the spreadsheet recalculate the time.

The spreadsheet uses the familiar formula for interest to find how much of the payment applies to interest to date but also includes how much of the payment applies to the principal and what the loan balance is after each payment. Simply using the formulas, however, creates error messages on the spreadsheet if no payment is yet entered, so we will use the ISBLANK function to enter empty cells until a payment is recorded. The formulas we will use are:

Interest Paid =IF(ISBLANK(Payment),then "",
*otherwise Loan Balance*Rate*((Payment Date - Loan Date)/365))*

Principal Paid =IF(ISBLANK(Payment),then "",otherwise Payment Amount - Interest Paid)

Current Loan Balance =IF(ISBLANK(Payment),then "",
otherwise Previous Loan Balance - Principal Paid)

Exercise 1

On April 1, Mary Carrington Interiors took out a 90-day note for $1,500 bearing interest at 10%. On May 1, Carrington paid $700 toward the obligation, and on May 16, she paid another $300. How much remains to be paid on the due date?

1. Open a new spreadsheet and enter the title of this portion of the spreadsheet, **Partial Payment of a Simple Interest Note**.
2. Skip a row and in the first column enter the label **Principal Amount** and three columns over in the same row enter **Loan Date**.
3. Below the Principal Amount label enter **Annual Interest Rate** and three columns across from that enter **Due Date**.
4. Enter the principal, interest rate, loan date, and the due date.
5. Skip a row and enter the column labels: **Payment Date, Payment Amount, Interest Paid, Principal Paid, Loan Balance**.
6. Make the first Payment Date equal to the loan date, which in our example is =E3.
7. Under the Loan Balance, enter the reference to the principal amount. In our example this is =B3.
8. Under the Payment Date, in the next row enter the first payment's date and next to it the first payment amount from the problem statement.
9. Use an absolute reference to the interest rate and enter the formula for the interest paid. In our example this is =IF(ISBLANK(B8),"",E7*$B$4*((A8-A7)/365)).
10. Enter the formula to find how much of the payment will go toward the principal. In our example it is =IF(ISBLANK(B8),"",B8-C8).
11. Enter the formula for the loan balance. In our example it is =IF(ISBLANK(B8),"",E7-D8).
12. Copy the formulas down the column and enter the second payment information.
13. For the final payment data enter the due date for the final payment date and enter an estimate for the Payment Amount that exceeds the previous Loan Balance. Adjust this amount until the final loan balance becomes $0.00.
14. Select the range of cells used plus an extra row with the PAYSIMPLE name.
15. Format the spreadsheet as shown in Figure 6-1 and save the spreadsheet as PAYMENTS.

The final payment on the Mary Carrington Interiors' note will be $522.02.

A1 | Partial Payment of a Simple Interest Note

	A	B	C	D	E	F
1	Partial Payment of a Simple Interest Note					
2						
3	Principal Amount	$ 1,500.00		Loan Date	4/1/96	
4	Annual Interest Rate	10%		Due Date	6/30/96	
5						
6	Payment Date	Payment Amount	Interest Paid	Principal Paid	Loan Balance	
7	4/1/96				$1,500.00	
8	5/1/96	$ 700.00	$ 12.33	$ 687.67	$ 812.33	
9	5/16/96	$ 300.00	$ 3.34	$ 296.66	$ 515.67	
10	6/30/96	$ 522.02	$ 6.36	$ 515.66	$ 0.00	
11						

Figure 6-1: With two payments made, this simple interest note will call for $522.02 remaining on a loan of $1,500.

Charge It!

Credit has always been a part of the American lifestyle, but the buying habits of Americans were changed drastically with the invention of the Installment plan. Over the years many people have been greatly deceived by the ease with which they could buy things even though they had no money.

As you work through the next three exercises you will see the United States rule at work as well as the Truth in Lending Law (Regulation Z). The Truth in Lending Law now makes every lender-to-be advertise a standard rate of interest, the APR (Annual Percentage Rate).

Monthly Installment Plans

Different arrangements can be made for fulfilling your obligation to a store or lending institution. While revolving charges are now the rule, consumers may find it advisable to "pay off" their debts by setting a payment schedule for themselves. A goal for paying a set amount per month and not purchasing more or borrowing more until they are free from the original loan can help to manage debt overall. The amount must be sufficient to cover interest and reduce the principal at the same time.

Steady payments of equal amounts will eventually reduce the amount due to something below the standard payment but still something will be owed. To control for this in our spreadsheet we will use an IF function that computes the last payment as the sum of the balance due and its interest.

Exercise 2

Mrs. Cohen's charge account at Butler-Stohr has a balance of $100. Interest of 1.5% per month is charged on unpaid customer accounts. Mrs. Cohen has decided to pay off her account by making monthly payments of $25. How much of each payment is interest and how much will apply toward her account? What is the total amount Mrs. Cohen will pay and how much of this is interest?

1. Open the PAYMENTS spreadsheet and go to I100 or some area far away from the Simple Interest Loan area.

Tip: *If you have a multiple-worksheet program, you can make a worksheet for each type of partial payment plan.*

2. Enter the title **Installment Plan with Monthly Rate**.
3. Skip a row and enter two labels and leave two cells between the labels: ***Account:, blank column, blank column, Monthly Payment.***
4. Fill in Mrs. Cohen as the Account and the Monthly Payment.
5. Skip a row and enter the column labels. Notice that we use the monthly rate's amount as both a label and a reference amount. Enter: ***Payment Number, Balance Due, 1.5%, Monthly Payment, Payment toward Balance Due, Adjusted Balance Due***
6. Enter the starting balance under the Balance Due column and in the first row.
7. Enter the formula for the monthly interest in the 1.5% column and row 1: Balance Due*1.5%, which in our example is =J105*$K$104.
8. Enter an IF function to control the monthly payment. =IF(Balance Due>Monthly Payment,then Monthly Payment,otherwise Balance Due+Interest), in our example =IF(J105>$N$102,$N$102,J105+K105).
9. In the Adjusted Balance Due column enter the formula that subtracts the Payment toward Balance Due from the Balance Due. In our example it is =J105-M105.
10. In the second payment's row, have the Balance Due cell equal row 1's Adjusted Balance Due. In Figure 6-2 this is =N105.
11. Copy the amounts and formulas down to row 5.

12. Enter SUM functions to total the 1.5%, Monthly Payment and Payment Toward Balance Due columns.
13. Select this area of cells and name it the INSTALL range.
14. Format the spreadsheet as you can see in Figure 6-2 and save the spreadsheet.

Your work should look like Figure 6-2.

I100 | Installment Plan with Monthly Rate

	I	J	K	L	M	N	O
100		**Installment Plan with Monthly Rate**					
101							
102	***Account:***	Mrs. Cohen		***Monthly Payment***		$ 25.00	
103							
104	***Payment Number***	***Balance Due***	***1.5%***	***Monthly Payment***	***Payment toward Balance Due***	***Adjusted Balance Due***	
105	1	$ 100.00	$ 1.50	$ 25.00	$ 23.50	$ 76.50	
106	2	$ 76.50	$ 1.15	$ 25.00	$ 23.85	$ 52.65	
107	3	$ 52.65	$ 0.79	$ 25.00	$ 24.21	$ 28.44	
108	4	$ 28.44	$ 0.43	$ 25.00	$ 24.57	$ 3.86	
109	5	$ 3.86	$ 0.06	$ 3.92	$ 3.86	$ -	
110	***Totals:***		$ 3.92	$ 103.92	$ 100.00		
111							

Figure 6-2: This portion of the spreadsheet shows how patient and consistent payments can eliminate a debt.

Revolving Credit

Many consumers have found it convenient and sometimes profitable to make their purchases on a revolving credit basis. In this way they can take advantage of sales and the availability of special products when they do not have ready cash. Many, however, find they spend too easily when given this "blank check," as if it will never need to be paid. They end up paying much more than if they had waited, saved, and paid with cash.

With revolving charges interest is paid on an average balance which takes into account the time between payments and the previous billing date as well as the time between the payment and the present billing date. It answers, "What was the average amount owed throughout the 30-day billing period per day before and after monthly payments?" The average balance does not include new debts incurred during the period. These are added in on the billing date.

COMPUTE THE AVERAGE DAILY BALANCE:

1. Find the number of days from the previous billing to the payment (this will equal the date of the payment day if the billing cycle is always at the end of the month) = *A*
2. Find the number of days from the payment to the new billing date (this will be 30 - given a 30-day billing cycle) = *B*
3. Multiply A times the Previous Balance. = *C*
4. Multiply B times the Adjusted Balance = *D.*
5. Add C + D and divide by the number of days in the billing cycle: *(C + D)/Number of days in Month,* and format it as currency.

For the spreadsheet we will use the DAY function to return the number for a day of the month. So the formula for the Average Daily Balance becomes:

*Average Daily Balance = ((DAY(Previous Balance Date)*Previous Balance)+((DAY(Current Date)-DAY(Previous Balance Date))*Adjusted Balance))/DAY(Current Date).*

Exercise 3

Mr. Boyer's charge account at Wymer's starts with a $50 balance at the beginning of a 6-month period. Payments and additional charges occurred as per Table 6-1 below. Build a spreadsheet that accounts for this activity:

Table 6-1

Month	Day	Activity	Amount
1	1/15	Opening Balance	$ 50.00
	1/15	Payment	$ 10.00
	1/20	Purchase	$ 25.00
2	2/5	Purchase	$ 30.00
	2/10	Payment	$ 10.00
	2/14	Purchase	$ 20.00
3	3/20	Payment	$ 20.00
	3/22	Purchase	$ 15.00
	3/28	Purchase	$ 10.00
4	4/10	Payment	$ 20.00
	4/18	Purchase	$ 15.00
5	5/8	Purchase	$ 25.00
	5/15	Payment	$ 20.00
	5/20	Purchase	$ 5.00
6	6/2	Purchase	$ 24.00
	6/12	Payment	$ 20.00
	6/22	Purchase	$ 15.00

Enter the Labels:

1. Open the PAYMENTS spreadsheet and go to a new area well away from the other two examples.
2. Enter the title, **Revolving Charge Account.**
3. Skip a row and enter ***Account:*** and next to it enter Mr. Boyer.
4. On the next row down enter: *blank column, blank column,* ***Revolving Account,*** *blank column, blank column, blank column, blank column,* ***Current Account.***
5. Skip another row and enter the column labels ***Mo., Date, Prev. Bal., Paymt., Adj. Bal., Avg. Bal., 2.0%, Purchases, Cur. Bal.***

Enter the First Month's Data and Formulas:

1. Enter 1 in the first row of the Mo column and January 15 as the first account activity day of the first month.
2. Enter the beginning balance, the payment and the formula for the adjusted balance (Balance - Payment). The Adjusted Balance formula in our example is =R206-S206.

3. Even though some of these cells are empty for this row, enter the formula for the current balance which will subtract the Payment, and add the Interest (2%) and purchase amounts. In the example the formula is =R206-S206+W206+V206.

4. Copy the Current Balance formula down to the next row and edit it to reference the previous current balance rather than a previous balance. In our example (Figure 6-3) the formula is =X206-S207+W207+V207.

5. Enter the activity for the 20th and copy the Current Balance formula down to the 20th's row.

6. Enter the end-of-month date as January 30 in the next date column and enter the formula for the Average Balance, which in our example is =((DAY(Q206)*R206)+((DAY(Q208)-DAY(Q206))*T206))/DAY(Q208).

7. Enter the formula for the Interest on the average daily balance, multiplying the 2% cell (an absolute reference) times the Average Balance. In our example this is =U208*$V$205.

8. Format the 1/31 row with a line below it and copy the current balance formula down to it.

Enter the Succeeding Months' Data and Formulas:

1. Enter the activity for 2/5 and copy the current Balance formula down as far as you judge will be necessary (see Figure 6-3).

2. For the Previous Balance column and 2/10, reference the current balance for 1/31. In the example this is =X208.

3. Enter the payment and copy the Adjusted Balance formula from above.

4. Enter the activity for the 2/14.

5. Enter 2/29 (our example uses a leap year date otherwise the last day of February is 28) and enter the formula for the Average Balance (in this case it references the correct cells but you may need to edit for the succeeding months).

6. Copy the formula for interest to the Interest cell.

7. Repeat the above until you complete the entries for all the activity in Table 6-1.

Total and Format the Spreadsheet:

1. Skip a row for future expansion and enter the **Totals** label and enter SUM functions for the Payment column, Interest column, and Purchase column.

2. Select the range of cells used and give it the REVOLVING name.

3. Format the spreadsheet as you can see in Figure 6-3 and save the spreadsheet.

Your work should resemble Figure 6-3.

P200 Revolving Charge Account

	P	Q	R	S	T	U	V	W	X
200				Revolving Charge Account					
201									
202	Account:	Mr. Boyer							
203			Revolving Account					Current Account	
204									
205	Mo.	Date	Prev. Bal.	Paymt.	Adj. Bal.	Avg. Bal.	2.0%	Purchases	Cur. Bal.
206	1	1/15/96	$ 50.00	$ 10.00	$ 40.00				$ 40.00
207		1/20/96						$ 25.00	$ 65.00
208		1/31/96				$ 44.84	$ 0.90		$ 65.90
209	2	2/5/96						$ 30.00	$ 95.90
210		2/10/96	$ 65.90	$ 10.00	$ 55.90				$ 85.90
211		2/14/96						$ 20.00	$ 105.90
212		2/29/96				$ 59.35	$ 1.19		$ 107.08
213	3	3/20/96	$ 107.08	$ 20.00	$ 87.08				$ 87.08
214		3/22/96						$ 15.00	$ 102.08
215		3/28/96						$ 10.00	$ 112.08
216		3/31/96				$ 99.99	$ 2.00		$ 114.08
217	4	4/10/96	$ 114.08	$ 20.00	$ 94.08				$ 94.08
218		4/18/96						$ 15.00	$ 109.08
219		4/30/96				$ 100.75	$ 2.02		$ 111.10
220	5	5/8/96						$ 25.00	$ 136.10
221		5/15/96	$ 111.10	$ 20.00	$ 91.10				$ 116.10
222		5/20/96						$ 5.00	$ 121.10
223		5/31/96				$ 100.78	$ 2.02		$ 123.11
224	6	6/2/96						$ 24.00	$ 147.11
225		6/12/96	$ 123.11	$ 20.00	$ 103.11				$ 127.11
226		6/22/96						$ 15.00	$ 142.11
227		6/30/96				$ 111.11	$ 2.22		$ 144.34
228									
229		Totals		$ 100.00			$ 10.34	$ 184.00	
230									

Figure 6-3: Revolving charges challenge you to keep track of early payments as well as increased debt.

Set Payment Plans

You can also get credit with agreed upon payment plans. In this case the partial payments are prearranged. You can find this arrangement with purchases, which are called Finance Plans. Financial plans usually include a down payment and a carrying charge or Finance charge.

Finance plans have a harder time translating their interest rates into the standard Annual Percentage Rate, which all lenders must now disclose so the consumer can compare financing deals with a standard rate. These values come from tables published by the U.S. government. You can calculate an equivalent figure called the Effective Rate.

To calculate the effective rate use the formula:

*Effective Rate = (2(Time in Months * Finance Charge Rate))/(Time in Months + 1)*

Exercise 4

Gerald Wood bought a new color television that sold for $675 cash. He received a $35 trade-in allowance on his old set and also paid $40 cash. The store arranged a 2-year payment plan, including an 8% finance charge. What is Wood's monthly payment? What will be the total cost of the TV? How much extra will he pay for the convenience of installment buying? What annual percentage rate must be disclosed under the Truth in Lending Law?

1. Open spreadsheet PAYMENTS.
2. Go to an area well away from the other examples enter the title **Payment Plans - Financing** and the labels down the column: *blank row,* **Account,** *blank column,* **Cash**

Price, Down Payment Rate, Down Payment, Outstanding Balance, *blank column,* **Finance Charge Rate, Time in Months, Time in Years, Finance Charge, Total of Payments,** *blank column,* **No of Payments, Monthly Payment,** *blank column,* **Total Cost,** *blank column, blank column,* **Effective Rate or APR.**

3. Put the Cash Price and other currency figures in the second column to the right of the labels (see Figure 6-4). Enter the known values from the problem statement above. What was the total down payment?
4. Enter the formula for the Outstanding balance: Cash price - Down Payment, which in our example is =AC304-AC306.
5. Enter the formula for Time in Years: Time in months/12, which in our example is =AB310/12.
6. Enter the formula for the Finance Charge: Principal*Rate*Time or Cash Price*Down Payment*Time in Years, which in our example is =AC307*AB309*AB311.
7. Find the Total Payments: Outstanding Balance + Finance Charge, which in our example is =AC307+AC312.
8. The payments are monthly in this case, so enter the number of payments as equal to the Time in months. In our example this is =AB310.
9. Compute the payment amount: Total Payments/Number of payments, which in our example is =AC313/AB315.
10. Enter the Total Cost formula: Down Payment + Total Payments, which in our example is =AC306+AC313.
11. Find the Effective Interest Rate or APR formula. In our example the formula is =(2*(AB315)*(AB309))/(AB315+1).
12. Select the range of cells used and give it the PPFINANCE name.
13. Format the spreadsheet as you see in Figure 6-4 and save the spreadsheet.

SetPayments | Payment Plans - Financing

	AA	AB	AC	AD	AE	AF
300	**Payment Plans**	**Financing**				
301						
302	**Account**	Gerald Wood		Note: Down payments can be determined in different ways: a percentage, straight amount, trade-ins, and the like.		
303						
304	**Cash Price**		$ 675.00			
305	**Down Payment Rate**					
306	**Down Payment**		$ 75.00			
307	**Outstanding Balance**		$ 600.00			
308						
309	**Finance Charge Rate**	8%				
310	**Time in Months**	24				
311	**Time in Years**	2				
312	**Finance Charge**		$ 96.00			
313	**Total of Payments**		$ 696.00			
314						
315	**No of Payments**	24				
316	**Monthly Payment**		$ 29.00			
317						
318	**Total Cost**		$ 771.00			
319						
320						
321						
322	**Effective Rate or APR**	15.36%				
323						
324						

Figure 6-4: The cost of financing a television with set monthly payments. The Finance Charge Rate of 8% represents an annual interest rate of 15.36%.

Early Payment of Installment Loans and the Rule of 78

With installment loans a borrower has use of 100% of the loan the first month and less each month following. The lender risks more money during the early part of the loan than later and so is allowed to collect more of the interest early on than later. This is a problem when borrowers pay off their loans earlier than planned and therefore deserve a refund of the interest portion of their final payment. The lender does not owe an averaged amount of interest as if the interest were the same each month.

What the lender must return is based on the Rule of 78. This rule is named after a one-year (12-month) loan. If you add 12+11+10+9+8+7+6+5+4+3+2+1 the sum is 78. This number, 78, thus represents 100% of the interest. The lender gets 12/78 or 15.4% of the interest in the first month of the loan and, conversely, owes a borrower who pays back his loan in the eleventh month 1/78 or 1.3% of the interest.

The formula for computing the Rule of 78 depends on finding the fractional portion of the time remaining represented by a numerator and denominator. The numerator is the sum-of-years'-digits for the remaining months while the denominator is the sum-of-years'-digits for the entire loan period. The sum-of-years'-digits is also used as a way to compute depreciation, which we will cover in a later chapter.

The numerator then is $a = (n*(n+1))/2$ where n is the number of remaining payments.

The denominator is $b = (Y(Y+1))/2$ where Y is the number of months in the loan period. For one year this is (12*(12+1))/2 = 78.

The Interest a borrower saves by paying early is the quotient of a/b multiplied by the total interest: I*(a/b).

Exercise 5

Mason Jansen borrowed $600 at 13% simple interest, to be repaid in 12 monthly payments. When making his eighth payment, Mr. Jansen wants to pay his remaining balance. How much interest will Jansen save? How much more does he have to pay?

1. Go to another area clear of all the other multiple payment plan work areas in the PAYMENTS spreadsheet.
2. Enter the title **Installment Plan Prepayments** and the row labels, *blank column,* **Name,** *blank column,* **Principal Amount, Annual Interest Rate, Time in Months, Time in Years, Total Interest, Maturity Value,** *blank column,* **No of Payments, Payment Amount,** *blank column,* **No Payments Made, No Payments Remaining, Numerator for Rule, Denominator for Rule,** *blank column,* **Total of Remaining Payments, Interest Saved, Balance Due.**
3. Enter the known data and format the cells appropriately, separating the currency from the other data as done above.
4. Enter the formulas for Time in Years, Interest, and Maturity Value. In our example they are =AG406/12, =AH404*AG405*AG407, and =AH404+AH408.
5. The Number of Payments equals the Time in Months and the Payment Amount is the Maturity Value divided by the Number of Payments. In our example they are 12 and =AH409/AG411.
6. The Number of Payments made includes this final payment. In our example it is 8.
7. The Number of payments remaining is the Number of payments minus the Number of payments made. In our example it is =AG411-AG414.

8. Enter the formula for the Rule of 78's numerator: (Number of payments remaining*(Number of payments remaining +1)) / 2, which in our example is =AG415*(AG415+1)/(2).

9. Enter the formula for the Rule of 78's numerator (do so even though you know 78 is always the denominator for a one-year loan so you can use this template for shorter and longer loan periods): (Number of payments*(Number of payments =1)) / 2, which in our example is =AG411*(AG411+1)/(2).

10. Find the Total of the Remaining Payments: No Payments Remaining * Payment Amount, which in our example is =AH412*AG415.

11. Find the Interest Saved: (Numerator for rule/Denominator for rule) * Interest, which in our example is =(AG416/AG417)*AH408.

12. Find the Balance Due: Total of remaining payments - Interest saved, which in our example is =AH419-AH420.

13. Name the area used as RULE78, format the area as seen in Figure 6-5, and save the spreadsheet.

Rule78 | Installment Plan Prepayments

	AF	AG	AH	AI
400	**Installment Plan Prepayments**			
401				
402	**Name**	Mason Jansen		
403				
404	**Principal Amount**		$ 600.00	
405	**Annual Interest Rate**	13.0%		
406	**Time in Months**	12		
407	**Time in Years**	1		
408	**Total Interest**		$ 78.00	
409	**Maturity Value**		$ 678.00	
410				
411	**No of Payments**	12		
412	**Payment Amount**		$ 56.50	
413				
414	**No Payments Made**	8		
415	**No Payments Remaining**	4		
416	**Numerator for Rule**	10		
417	**Denominator for Rule**	78		
418				
419	**Total of Remaining Payments**		$ 226.00	
420	**Interest Saved**		$ 10.00	
421	**Balance Due**		$ 216.00	
422				

Figure 6-5: The Rule of 78 finds how much interest is due the borrower who pays off a loan early.

Exercise 6

Create a template file from this worksheet.

Application Exercises

Use the PAYMENTS spreadsheet to find the answers to the following Multiple Payment Plan problems.

1. On March 3, Andrew Myer signed a $3,000 note with interest at 8% for 270 days (add the number of days to the loan date to find the due date). Myer made payments of $780 on July 1 and $846 on September 29. How much will he owe on the due date?

2. Griffin Parts Co. borrowed $5,000 by signing a 9% note for 240 days on March 6. A payment of $625 was made toward the note on March 26, and $2,166 was paid on May 25. Find the balance due on the day the note matures.

3. On February 20, 1996, Albert Rogers borrowed $2,400 by signing a 9%, 240-day note. He made a $536 partial payment on April 21 and paid $538 on July 10. Determine the amount he still owes at the end of the loan period.

4. Mitchell Candy Co. borrowed $4,000 by signing a 9% note dated June 15 for 150 days. The firm made a $520 partial payment on July 5 and another payment of $1,070 on September 23. what payment will be required when the note is due?

5. Everett Department Store charges 1% per month interest on adjusted charge account balances. Lori Beulow charged a $70 dress and agreed to pay $15 monthly toward her account. How much interest will she pay?

6. Anita Alfred's account balance at Blanchard Department Store is $500. The store charges 1.5% per month interest on adjusted charge account balances. If she agrees to pay $60 monthly toward her account, how much interest will she pay?

7. Purchases are listed below for Barry Stewart's revolving charge account with Maxwell's, where a 1% interest charge is computed on the account's average daily balance. No interest is charged on new purchases made during the 30-day month. Maxwell's requires a $10 payment when the previous month's ending balance is under $100 and $15 for balances over $100. Find Barry's total payments, total interest, and final balance if he only pays the minimum each month. (***Tip**: Create an IF statement to determine the appropriate minimal payment*).

Month	Day	Activity	Amount
1	3/12	Beginning Balance	$70
	12	Payment	
	17	Purchase	$30
	21	Purchase	$10
2	10	Payment	
	26	Purchase	$18
3	15	Payment	
4	5	Purchase	$25
	9	Purchase	$20
	20	Payment	

8. Van Lamar's charge account reflects the following purchases. Interest at 1% is computed on his average daily balance (but not on current purchases); a $15 payment is required when the preceding month's ending balance is under $100, and $20 for balances exceeding $100. Assuming minimum payments, find the total payments, total interest, and ending balance.

Month	Date of Payment	Amount	Activity
1		$90	Beginning Balance
	10		Payment
	15	$20	Purchase
2	5		Payment
	10	$30	Purchase
3	13	$25	Purchase
	20		Payment
	24	$25	Purchase
	25	$10	Purchase

Complete an analysis of each of the following finance loan plans (problems 9 - 15) showing the payments and total cost for each.

9. A food freezer sells for $810 cash or a one-third down payment, a finance charge of $60, and the balance in 24 monthly installments.

10. A camcorder is priced at $900 cash or a one-quarter down payment, a finance charge of $45, and the balance in 36 monthly installments.

11. A bed and dresser set is marked at $450. On the time-payment plan, a 20% down payment is required and the 18 monthly payments include a $54 finance charge.

12. A fax machine is marked $470. The time-payment plan requires a 20% down payment, 24 monthly payments, and a finance charge of $50.

13. A diamond ring is priced at $1,200. The time-payment plan requires a 10% down payment and 18 monthly payments. There is a 15% finance charge. What is the effective rate?

14. A couple purchased a washer and dryer costing $800. The appliances may be purchased for a 25% down payment and 24 monthly payments. The finance charge rate is 9%.

15. A loan company computes 15% nominal interest on a loan of $200 to be repaid in 4 monthly payments.

Show the computations on the following as you determine the Interest Saved and the Balance Due back to the borrower

16. Jason Goode took out a $3,000 car loan at 12% to be repaid in 24 monthly payments. Jason pays off the remaining balance when he makes his sixteenth payment.

17. Josh Jeffers bought furniture for $2,000 under a finance plan of 12 monthly payments at 7% simple interest. Josh makes his 8th payment and pays off his remaining balance.

18. Amanda Pope purchased a refrigerator with a cash price of $900. She purchased on an installment plan with payments over 9 months at 12% simple interest. She pays it off on the fifth payment.

19. Edwin King borrowed $1,500 at 8% simple interest to be repaid in 10 monthly payments. He pays it all on his sixth payment.

20. Jeanne Shay is buying a $2,700 personal computer by making 18 monthly payments at 8% simple interest. She pays it all on the thirteenth payment.

21, Tony Mendoza is buying a $700 CD/stereo system by making 24 monthly payments at 10% simple interest. He pays all after the 18th payment.

22. Ginger Flanders borrowed $6,000 for investing in a gift shop by signing an 8% simple interest note requiring 24 monthly payments. She pays all at the 18th payment.

23. Chris Watson purchased a $1,800 health club membership by making 24 monthly payments at 9% simple interest. He pays off the balance with his fourteenth payment.

III. Savings and Investments

7. Savings

A savings account kept by an individual or company is usually an investment that earns compound interest. Compound interest is added to the principal periodically. This chapter will deal with saving a lump sum of money over time.

Objectives

By working through and completing the exercises below, you will be able to:

- Create a template and show the details of a long-term savings plan using compound interest.
- Create a template that quickly shows the future value of savings for partial years.
- Prepare a savings account ledger.
- Create a template to find how much to deposit now to have X amount in X number of years.

Compound Interest and Future Value

Compound interest earns more for the investor than simple interest because the borrower or lending institution pays interest on interest as well as the principal. When offered compound interest, you can use a spreadsheet to answer, "How much will I have if I deposit X amount today and it earns compound interest for X number of years?" In this case you are looking for the Future Value of your investment.

Future Value–Long Method

Computing the Future Value of an investment in detail helps you study your investment as it progresses over the years. To create it you will show the interest earned and end-of-period principals as interest is added to your initial investment over the years. However, first you will find the end-of-period principals and then find the interest for each period using the formulas:

*End-of-Period Principal = Beginning Principal * (1 + Period Interest Rate)*

Interest = Previous End-of-Year Principal - Beginning Principal

Exercise 1

Richard and Ann have saved $10,000. They now qualify to invest it in a special savings plan that pays 8% annually and compounded semiannually. How much interest will they earn each period and how much will their investment have grown in three years?

1. Start a new spreadsheet and save it as SAVINGS.
2. Create the headings in Table 7-1.

Table 7-1

Future Value–Long Method			
Savings Account			
Beginning Principal			
Annual Interest Rate			
Period Interest Rate			
Period	**Beginning Principal**	**Interest**	**End-of-Period Principal**

3. Fill in the Beginning Principal, Annual Interest Rate, and Period Interest Rate from the problem statement. Name the Period Interest Rate's cell the PeriodRate.

Note: *You should be able to easily enter the formula for any given Period Interest Rate by dividing the Annual Interest Rate by the appropriate portion of the year (for example, in this case, semiannual means twice a year so you must divide by two).*

4. Beneath the Period column heading create a series of numbers incrementing by one. For this example you will need 6 periods.

5. Make the first entry under the Beginning Principal column equal to the Beginning Principal row above it. In our example this is =C4.

6. Skip the Interest column and for the first end of period principal enter the formula for this: =First Principal * (1 + Period Interest Rate). This is 104% of the principal. In our example this is =B8*(1+PeriodRate).

7. The interest is the first End of Period Principal minus the principal. In our example this is =D8-B8.

8. Make the second Beginning Principal equal to the first End of Period Principal: =D8.

9. Copy the formulas down to the final period.

10. Name the range of used cells FVLONG so you can quickly return to it.

11. Format your work as you can see in Figure 7-1 and save the spreadsheet.

Your spreadsheet should look like Figure 7-1.

A1 | Future Value - Long Method

	A	B	C	D	E
1		**Future Value - Long Method**			
2		**Savings Account**			
3					
4	**Begining Principal**		$10,000.00		
5	**Annual Rate**		8%		
6	**Period Interest Rate**		4%		
7	***Period***	***Beginning Principal***	***Interest***	***End of period principal***	
8	1	$ 10,000.00	$ 400.00	$ 10,400.00	
9	2	$ 10,400.00	$ 416.00	$ 10,816.00	
10	3	$ 10,816.00	$ 432.64	$ 11,248.64	
11	4	$ 11,248.64	$ 449.95	$ 11,698.59	
12	5	$ 11,698.59	$ 467.94	$ 12,166.53	
13	6	$ 12,166.53	$ 486.66	$ 12,653.19	
14					

*Figure 7-1: With this Future Value template you can compute interest earned monthly, quarterly, semi-annually, or annually. After three years Richard and Ann's investment will have earned **$2,653.19** in interest.*

Exercise 2

Phylis Dwier manages The Hartford Corporation's employees' retirement fund. She is considering placing a portion of the $45,000,000 fund in a long-term savings plan yielding 5% interest compounded quarterly. If she does, the money will not be available for withdrawal for 10 years. She will need 30% of the funds before that time to meet her present payments and fees and wants to know how much the corporation will make if she commits the rest of the money to the savings plan. She will need to compose a report showing the growth of the fund over the years. Use the Future Value Long Method spreadsheet to show the details. Write the report for Phylis. Remember to open your template and save it immediately as the Phylis spreadsheet.

Future Value–Short Method

The Future Value Long Method takes more time and space than is needed for just finding the future value of a savings deposit if all things stay the same over time. With interest compounded daily you will also find the size of your spreadsheet hard to manage. If you simply want to find one specific year's future value you will find the short method much more direct. You can use a formula or the Future Value FV function. It computes the same final information that the long method presents by using the formula:

Future Value = Principal(1 + Period Rate)^Number of periods*

OR

Future Value =FV(Interest Rate/Periods per year,
*Number of years * Periods per Year, 0, 0, Beginning Balance)*

Exercise 3

Use Richard and Ann's problem above to find the future value and total interest after 10 years of saving.

1. Start a new portion of the SAVINGS spreadsheet, leaving room for growth below the Long Method portion. The example in Figure 7-2 begins in cell G55.
2. Enter the title and labels **Future Value–Short Method,** *blank row,* **Beginning Balance,** *blank row,* **Annual Rate, No Periods per Year, Period Rate,** *blank row,* **No Years, Total No Periods,** *blank row,* **Future Value,** *blank row,* **Total Interest**
3. Enter the Principal, Annual Rate, 2 for the No Periods per Year, and the No of Years.
4. Enter the formula for the period rate. =Annual Rate/No Periods per Year, which in our example is **=H59/H60.**
5. Enter the formula for number of periods. No Periods per Year * No Years, which in our example is **=H60*H63.**
6. Enter the formula for Future Value. In our example it is **=H57*((1+H61)^H64)** OR, using the FV function, it is **=FV(H59/H60,H60*H63,0,-H57).**
7. Enter a formula to subtract the principal from the Future Value to find the Total Interest. In our example this is **=H66-H57.**
8. Select the range of cells used to find Future Value using the short method and name this range FVSHORT.
9. Format your work as you see in Figure 7-2 and save the spreadsheet.

Your work should look like Figure 7-2.

G55		Future Value - Short Meth	
	G	H	I
55	Future Value - Short Method		
56			
57	Beginning Balance	$ 10,000.00	
58			
59	Annual Interest Rate	8.0%	
60	No Periods per Year	2	
61	Period Interest Rate	4.0%	
62			
63	No Years	3	
64	Total No Periods	6	
65			
66	Future Value	$12,653.19	
67			
68	Total Interest	$ 2,653.19	
69			

Figure 7-2: The Future Value–Short Method is the direct route to seeing how your savings can grow.

Tip: *It is often the case that you will have more than one deal to take for your investment. To use your existing template, insert columns to the right of the Future Value ranges and copy the formulas to the right as needed. Entering different constant values will show you how your different options compare.*

Exercise 4

Use the information from the Hartford Corporation's employee retirement fund to compute the future value. Create a five-minute presentation on this single set of information to present to the retirement fund board of supervisors. Would a pie chart and line chart help you make your point?

Future Value for Partial Years and a Savings Account

You can use the Future Value–Short Method spreadsheet to find the value of a one-lump amount savings account. All you need to do is compute the fractional amount for the number of years.

A more common need is a savings account ledger similar to the check ledger you created in Chapter 2. It uses daily compounded interest and allows you to account for deposits and withdrawals. To do so you can create **Error! Reference source not found.** using the following formulas, which include the ISBLANK, IF, and Future Value or FV function.

Second and Succeeding Balances =IF(ISBLANK(Current Date),then "",otherwise FV(Interest Rate/365,Current Date - Previous Date,0,-Previous Balance)+Current Deposit - Current Withdrawal)

Interest =IF(ISBLANK(Date),then "", otherwise Current Balance - Previous Balance - Current Deposit Current Withdrawal)

Table 7-2

Compound Interest Savings Account				
Amount Invested		$10,000.00		
Annual Interest Rate		8%		
Date	***Deposit***	***Withdrawal***	***Interest***	***Balance***
1/1/96				$10,000.00
4/10/96			$ 221.57	$10,221.57

Exercise 5

Richard and Ann want to know the balance of their savings account each time they make a deposit. They put their $10,000 in a savings plan that pays daily compounded interest. They do so on January 1. Build Figure 7-3 applying the given formulas. In our example, the formulas are:

Second Balance =IF(ISBLANK(A8),then "",otherwise FV($C$4/365,A8-A7,0,-E7)+B8-C8)

Interest =IF(ISBLANK(A8),then "",otherwise E8-E7-B8+C8)

Enter a Deposit of $250 on April 4, and another of $1,000 on May 15. Name the range of cells used SAVEACCT.

Your spreadsheet should look like Figure 7-3.

A1 | Compound Interest Savings Account

	A	B	C	D	E	F
1	**Compound Interest Savings Account**					
2						
3	**Amount Invested**		$10,000.00			
4	**Annual Interest Rate**		8%			
5						
6	***Date***	***Deposit***	***Withdrawal***	***Interest***	***Balance***	
7	1/1/96				$10,000.00	
8	4/10/96	$ 250.00		$221.57	$10,471.57	
9	5/15/96	$1,000.00		$ 80.63	$11,552.20	
10						

Figure 7-3: A savings account ledger that accounts for daily compounded interest.

Exercise 6

Phylis Dwier has two other possible investment options for The Hartford Corporation employee's retirement fund. The first is the one given above, which is from the Citizen's Investment Corporation. The Ventura Savings and Loan offers 5.5% semi-annually, and the First National Bank offers 2% interest compounded daily. In addition, the board of supervisors will only allow tying up the money for two and a half years, in which case the Citizen's Investment Corporation will only offer a 3.75% rate. What is the best deal for the fund? Prepare a bar chart to illustrate your findings and write a report to the board of supervisors' chairperson explaining the options.

Present Value

Present value answers the question, "How much must I invest now to have X amount in X number of years?" Thus, it finds the lump sum deposit for a savings plan given a future goal. The formula for this is:

Present Value = Future Value / (1 + Period Rate)^No of Periods

OR

*Present Value =PV(Annual Interest Rate/No Periods per Year, No Periods per Year*No of Years,0,-Future Value)*

Exercise 7

Richard and Ann have set themselves a goal to save $40,000 in 20 years. How much of their $10,000 would they need to invest in savings at the 8% rate now to reach their goal?

1. Open the SAVINGS spreadsheet.
2. Copy the FVSHORT range to a new section of the spreadsheet.

3. Change the title to **Present Value**.
4. Change the Future Value label to **Present Value** and the Principal Amount label to **Future Value**.
5. Enter the $40,000 goal in the Future Value and 20 years in the No of Years value.
6. Enter the formula for Present Value. In our example it is =O140/((1+O146)^O149) or =PV(O141/O142,O142*O145,0,-O139)
7. Name this range of these used cells plus an extra row PRESENTVALUE.
8. Format the work as you see in Figure 7-4 and save the spreadsheet.

Your spreadsheet should look like Figure 7-4.

N137	Present Value	
N	**O**	**P**
137 **Present Value**		
138		
139 **Future Value**	$ 40,000.00	
140		
141 **Annual Interest Rate**	8.0%	
142 **No Periods per Year**	2	
143 **Period Interest Rate**	4.0%	
144		
145 **No Years**	20	
146 **Total No Periods**	40	
147		
148 **Present Value**	$8,331.56	
149		
150 **Total Interest**	$ 31,668.44	
151		

Figure 7-4: The present value of $8,331.56 will make $31,668 interest and a total of $40,000 in 20 years.

Exercise 8

The Hartford Corporation's Employee Retirement Fund board of supervisor's chairperson finds Phylis Dwier's projection of future needs far less than what she expects to earn in savings. The chairperson's study expects they will need $560,780,000 in the fund at the end of 10 years. Given the rates at her disposal in Exercise 6, how much will Phylis need to invest now to make sure she has this money available? Prepare a memorandum for the chairperson, including a chart.

Exercise 9

Make this spreadsheet into a template at this point.

Application Exercises

1. Find the interest on $800 at 13% annually, compounded daily, for 28 days.

2. Thayer Farm Trust made a farmer a loan of $1,200 at 16% for three years, compounded annually. Find the future value and the interest paid on the loan.

3. Maeola Killebrew invests $3,800 at 7%, compounded semiannually for two years. What is the future value of the investment and how much interest will she earn over the two-year period?

4. Carolyn Smith borrowed $6,300 at 8.5% for three years, compounded annually. What is the future value of the loan and how much interest will she pay on the loan?

5. First State Bank loaned Doug Morgan $2,000 for four years, compounded annually at 8%. How much interest was Doug required to pay on the loan?

6. A loan of $8,000 for two acres of woodland is compounded quarterly at 12% for five years. Find the future value and the interest.

7. Compute the future value and the interest on a loan of $10,500, compounded annually for four years at 10%.

8. Find the future value of an investment of $10,500 if it is invested for four years and compounded quarterly at 8%.

9. Ross Land has a loan of $8,500, compounded quarterly for four years at 6%. What is the future value of the loan?

10. Find the interest on $2,500 at 13.25%, compounded daily by Leader Financial Bank for 20 days.

11. How much interest is earned on a deposit of $1,500 at 12%, compounded daily for 30 days?

12. John McCormick has found a short-term investment opportunity. He can invest $8,000 at 12.25% interest for 15 days. How much interest will he earn on this investment if the interest is compounded daily?

13. The Absorbant Diaper Company will need $20,000 in ten years to buy a new diaper edging machine. How much must the firm invest at the present time if it receives 10% interest compounded annually?

14. Compute the amount of money that should be set aside today to ensure a future value of $2,500 in one year if the interest rate is 11% annually, compounded annually.

15. How much money should Linda Bryan set aside today to buy printing equipment in one year that costs $8,500? The current interest rate is 7.5% annually, compounded annually.

16. Ronnie Cox has just inherited $27,000. How much of this money should he set aside today to have $21,000 to pay cash for a Ventura Van, which he plans to purchase in one year? He can invest at 7.9% annually, compounded annually.

17. Shirley Riddle has just received a $10,000 gift from her mother. She plans to make a major renovation to her home and have enough left to invest for one year. She also plans to take a trip, which is projected to cost $6,999. Current interest rates are at 8.3% annually, compounded annually. How much does Shirley need to set aside today for her trip?

18. Rosa Burnett thinks she will need $2,000 in three years to make the down payment on a new car. How much must she invest today if she will receive 8% interest annually, compounded annually?

19. Calculate the amount of money that must be invested now at 6% annually, compounded quarterly, to obtain $1,500 in three years.

20. What sum must be invested now at 9%, compounded monthly to accumulate $8,000 in 6 years? How much interest will be earned?

21. Mr. Gudfather wants to provide a $10,000 wedding gift for his daughter. She is now 9 years old, and he would like the fund to be available by the time she is 20. He decides on an investment that pays 8% compounded semiannually. How large must the deposit be? How much of the final $10,000 will be interest?

22. Mr. Speedy has been financing his cars and realizes that this adds substantially to the cost. He decides to invest funds in advance so that he can buy his next car for cash and avoid the finance charge. He expects to need a new car in 5 years and to pay about $12,000 for it. He can put a sum of money in a safe investment at 5%, compounded quarterly or a more risky investment at 9%, compounded monthly. How much must he invest now to give him $12,000 in 5 years in each plan?

23. How much must be invested at 16.5%, compounded monthly, to amount to $20,000 in 7 years? How much interest will be earned?

24. Kate Richards promised her nephew that she would give him $3,000 when he graduates from high school in 4 years. If money is worth 9%, compounded quarterly, how much must she invest today so that the money will be available when he graduates?

25. EZ Loan Company loaned $500 at 8% annual interest, compounded quarterly for one year. Calculate the amount the loan company will earn in interest?

26. Mario Piazza was offered $900 now for one of his salon photographs or $1,400 in one year for the same photograph. Which would give Mr. Piazza a greater yield if he could invest the $900 for one year at 16%, compounded quarterly?

27. Alan Ruscyk can invest $10,000 at 8%, compounded quarterly for two years. Or he can invest the same amount at 8.2%, compounded annually for the same two years. If all other conditions (such as early withdrawal penalty and the like) are the same, which deal should he take?

28. An investment of $2,000 is made at the beginning of each year for three years. The investment is compounded annually at 8%. Find the future value and the interest at the end of three years.

29. Your company plans to enter several short-term financing agreements. It is your assignment to compute interest for monthly compounding at the annual rate of 8%. Construct an interest rate table showing the current principal and the interest earned for each of the twelve months for $1.00, compounded monthly.

30. One real estate sales technique is to encourage customers or clients to buy today, because the value of the property will probably increase during the next few years. If you receive the offer of, "Buy this lot today for $30,000. In two years, it will sell for $32,500," would it be a wise investment if, as an alternative, you could invest up to $30,000 at 12%, compounded annually? How much would you need to invest to earn the $32,500.

8. Annuities and Sinking Funds

Annuities are another way to save, and a sinking fund is another way to find how much to save to reach a goal. If you wish, creating the templates for these and the upcoming spreadsheets in this section would be a good time to use the multiple worksheet feature of your spreadsheet program. So you might consider using the SAVINGS spreadsheet and adding a new worksheet to it. If so, open the SAVINGS spreadsheet and label a worksheet Annuities-Sinking.

Objectives

By working through and completing the exercises below, you will be able to:

- Prepare a template to study the details of an Ordinary Annuity.
- Prepare a template to quickly show the Future Value of an Ordinary Annuity.
- Prepare a template to study the details of an Annuity Due.
- Prepare a template to quickly show the Future Value of an Annuity Due.
- Prepare a template to find the payment for a Sinking Fund savings plan.
- Prepare a template to find the present value of an Ordinary Annuity.

Annuities

An annuity is a savings plan to which the investor pays a regular amount following a regular payment schedule. There are two kinds of annuities: the Ordinary Annuity and an Annuity Due.

Investors in an Ordinary Annuity make their payments at the end of the payment period whereas investors with an Annuity Due make payments at the beginning of each payment period. Thus Ordinary Annuities do not pay interest on the first annuity payment. An Annuity Due does collect interest on the first annuity payment.

There are two ways to use the spreadsheet to compute annuities. The long method uses simple formulas to provide details on how the investment progresses over time. The short method uses the Future Value (FV) function to quickly answer the question of future value of an investment.

Ordinary Annuity Long Method

The long method for computing an Ordinary Annuity uses the interest formula you already know. You will simply create a running balance spreadsheet to show equal payments over time and the accumulated interest. This is very similar to the SAVEACCT range without the complication of withdrawals. Since the payment is at the end of the period, the interest is based only on the beginning balance:

Interest =Beginning Period Balance(Interest Rate/Periods per Year)*

Exercise 1

Your company wants to know what the future value might be for different amounts in the thousands invested in an ordinary annuity for three years. Create a table showing the per-thousand value of an ordinary annual annuity investment at 8% annual interest. To do so you will create the beginning of an annuities and sinking funds template.

1. Create a new spreadsheet called ANNUITY or a new worksheet in your multi-worksheet program.
2. Create the title and subtitle: **Future Value – Long Method,** Ordinary Annuity

3. Skip a row and enter the row labels: **Annuity Payment, Annual Interest Rate, Periods per Year**.
4. Skip a row and enter the column labels: **Period, Beginning Period Balance, Payment, Interest, Ending Period Balance**.
5. Enter the given values from the problem statement. In this case there is only one period per year.
6. In the first row in the Ending Balance column, enter 0.
7. Create a series of 1 to 4, beginning in the next row in the Period column. In our example this series begins in cell A10.
8. Next to this (in our example, in cell B10) make the Beginning Balance equal to the previous ending balance. In our example this is =E9.
9. Enter an absolute reference to the Annuity Payment amount for the first Payment amount. In our example the first payment =$C$4.
10. Enter the formula for interest, which in our example is =B10*($C$5/$C$6).
11. Enter a formula for the Ending Period Balance, =Beginning Period Balance + Payment + Interest, which in our example is =B10+C10+D10.
12. Copy the Beginning Period Balance, Payment, Interest, and Ending Period Balance formulas down the columns.
13. Select the range of cells used plus one extra row and name the range OAFVLONG.
14. Format the spreadsheet as you see in Figure 8-1 and save your spreadsheet.

Your spreadsheet should look like Figure 8-1.

A1 | Future Value - Long Method

	A	B	C	D	E	F
1			**Future Value - Long Method**			
2			Ordinary Annuity			
3						
4	**Annuity Payment**		$ 1,000.00			
5	**Annual Interest Rate**		8%			
6	**Periods per Year**		1			
7						
8	***Period***	***Beginning Period Balance***	***Payment***	***Interest***	***Ending Period Balance***	
9					$ -	
10	1	$ -	$ 1,000.00	$ -	$ 1,000.00	
11	2	$ 1,000.00	$ 1,000.00	$ 80.00	$ 2,080.00	
12	3	$ 2,080.00	$ 1,000.00	$ 166.40	$ 3,246.40	
13	4	$ 3,246.40	$ 1,000.00	$ 259.71	$ 4,506.11	
14						

Figure 8-1: Your company can now use the values in this table multiplied by whatever factor to find their future values (for example the future value of $4,000 annuity payments after 3 years would be 4 times the third year value).

Exercise 2

This is a challenge requiring you to design a version of the OAFVLONG worksheet.

The Vans currently own a home in Rhode Island. They wish to purchase a new home in 16 or 17 years without selling their current home. To save for the new down payment, they contribute $250 a quarter to an ordinary annuity for 16 years at 8% annual interest compounded semiannually. How much is each annuity payment? How much will they have for their down payment? As their financial advisor, prepare a report showing them how the money will increase over the years.

Tip: *Think about the amount paid per period rather than how many deposits the Vans make per year. Should they earn interest on every deposit?*

Annuity Due–Long Method

The long method for computing an Annuity Due resembles the Ordinary annuity; however you will see how the first Annuity Payment counts when computing interest. This calls for a different formula for computing the first and subsequent interest to include interest on the payment. These formulas are:

*Interest =(Beginning Period Balance + Payment) * (Annual Interest Rate / Number of Periods per Year)*

Exercise 3

What is the future value of a quarterly annuity due of $100 for 1 year at 10% annual interest compounded quarterly? Find the total investment and the total interest earned.

1. Go to a new area in the ANNUITY worksheet.
2. Copy the OAFVLONG range to this area.
3. Enter the data from the problem statement above.
4. Create a new series of Period numbers starting in the first row beneath the Period column label. (You need four quarterly periods.)
5. Enter an absolute reference to the Annuity Payment amount in this first row as well.
6. Enter the new formula for computing interest. In our example it is =(N104+O104)*($O$100/$O$101).
7. Enter the formula for the first Ending Period Balance. In our example it is =N104+O104+P104.
8. Copy the new Interest formula down the column.
9. Name the range of cells the ADFVLONG range.
10. Format the range as you see in Figure 8-2 and save the spreadsheet.

Your spreadsheet should look like Figure 8-2.

M96 | Future Value - Long Method

	M	N	O	P	Q	R
96		**Future Value - Long Method**				
97		Annuity Due				
98						
99	**Annuity Payment**		$ 100.00			
100	**Annual Interest Rate**		10%			
101	**Periods per Year**		4			
102						
103	***Period***	***Beginning Period Balance***	***Payment***	***Interest***	***Ending Period Balance***	
104	1	$ -	$ 100.00	$ 2.50	$ 102.50	
105	2	$ 102.50	$ 100.00	$ 5.06	$ 207.56	
106	3	$ 207.56	$ 100.00	$ 7.69	$ 315.25	
107	4	$ 315.25	$ 100.00	$ 10.38	$ 425.63	
108						

Figure 8-2: Because an annuity due is paid at the beginning of each period, the payment earns interest in the first period.

Exercise 4

Sarah Smith plans to invest a total of $8,000 over two years time at 8% annual interest. She is considering a quarterly ordinary annuity plan of $1,000 payments, compounded quarterly. She is also considering a semiannual ordinary annuity of $2,000 payments, compounded semiannually at 8%. Which would yield the most earnings? As her financial advisor, prepare a brief report explaining her options.

Ordinary Annuity and Annuity Due–Short Method

To compute the Future Value of an Ordinary Annuity in a single step you could use the complicated formula that resembles the Future Value you know from Chapter 7. But, as you may have noticed in Chapter 7, the FV function is really made for Annuities. This one-step method provides the information you are looking for without all the details of the long method and you can control for either an Ordinary Annuity or an Annuity Due with the formula:

Future Value =FV(Annual Interest Rate/Number Periods per Year,
*Number of Periods per Year*Number of Years,Annuity Payment,Annuity Type)*

Exercise 5

Find the future value of a semiannual ordinary annuity of $6,000 for 5 years at 12% annual interest compounded semiannually.

1. In a new section of the 0800 spreadsheet, create the title and row labels: **Future Value–Short Method**, *blank row*, **Annuity Payment**, *blank row*, **Annual Interest Rate**, *blank row*, **No Period Yr**, *blank row*, **No of Years, Annuity Type**, *blank row*, **Future Value**
2. Enter the given data.
3. Enter 0 for the Annuity Type and in the cell next to this enter 0=Ordinary Annuity; 1=Annuity Due.
4. Enter the FV function referencing the Annuity Type cell. In our example this is =FV(K52/K54,K54*K56,K50,,K57).

Note: *For an Excel spreadsheet, you must make the Annuity Payment a negative number to have the FV function return a positive amount. In this case the FV formula is* =FV(K52/K54,K54*K56,-K50,,K57).

5. What is the Annuity Due Future Value? (Enter 1 for the Annuity Type.)
6. Name the used range of cells AFVSHORT.
7. Format the range as you see in Figure 8-3 and save the spreadsheet.

Your spreadsheet should look like Figure 8-3.

J48	Future Value - Short Method				
	J	K	L	M	N
48	Future Value - Short Method				
49					
50	Annuity Payment	$ 6,000.00			
51					
52	Annual Interest Rate	12.0%			
53					
54	No Period Yr	2			
55					
56	No of Years	5			
57	Annuity Type	0	(0=Ordinary Annuity, 1=Annuity Due)		
58					
59	Future Value	$79,084.77			
60					

Figure 8-3: For a quick way to know the Future Value of an Annuity, you can use the short method for a given number of years.

Exercise 6

Building on Exercise 4, Sarah Smith has another possible offer. Using the short method for all her options, show her how a $325 annuity, due monthly for two years, at 7.5% annual interest compounded monthly, will compare to the previous deals. Amend your report to include this information.

Sinking Funds

A Sinking Fund is the payment computation for an Ordinary Annuity. Businesses and individuals use it to determine how much of an annuity to pay to reach a monetary goal in a set number of years. A business may do so to make good on bonds issued whereas you may do so to save for a special or large purchase.

The formula for finding the Sinking Fund Payment is:

*Payment = Future Value *(Period Rate/(1+ Period rate)^Number of periods - 1))*

You can use the PMT or Payment function to do the same task more easily:

*Payment = PMT(Rate/Periods per Year,Periods per Year*Years,0,Future Value)*

Exercise 7

Find the Annual Sinking Fund Payment required to accumulate $140,000 in 12 years at 8% annual interest.

1. Set up an new area of the ANNUITY worksheet
2. Copy the FVSHORT range to this new area.
3. Edit the Annuity Payment label to read **Future Value**.
4. Select the three cells for the Annuity Type row and delete them so that the cells below move up.
5. Enter the given data.
6. Enter the formula for the Annuity Payment (the sinking fund payment). In our example it is **=Z204*((Z206/Z208)/((1+(Z206/Z208))^(Z208*Z210)-1))** but you can use the PMT function. In our example it is **=PMT(Z206/Z208,Z208*Z210,0,-Z204)**.

7. Give this range the SINKFUNDPMT name.
8. Format the spreadsheet as you see in Figure 8-4 and save the spreadsheet.

Your spreadsheet should look like Figure 8-4.

Y202	Sinking Fund Payment		
Y	**Z**	**AA**	**AB**
202 **Sinking Fund Payment**			
203			
204 **Future Value**	$140,000.00		
205			
206 **Annual Interest Rate**	8.0%		
207			
208 **No Period Yr**	2		
209			
210 **Number of Years**	12		
211			
212 **Sinking Fund Paymei**	$ 3,582.16		
213			

Figure 8-4: The Sinking fund Payment is the What-If amount for meeting the goals of an Ordinary Annuity.

Exercise 8

The Bamboo Furniture Company manufactures rattan patio furniture. It has just purchased a machine for $13,500 to cut and glue the pieces of wood. The machine is expected to last five years. If the company wants to establish a sinking fund to replace this machine, what annual payments must be made if the annual interest rate is 8%?

Comparing a One Lump Sum Savings to an Ordinary Annuity

The Sinking Fund computation looks much like the present value of a compound interest savings plan. The Present Value of an Ordinary Annuity is a way to compare what lump sum payment in an equal interest rate and schedule would be required to earn the same payoff as an Ordinary Annuity plan.

The formula for the Present Value of annuity is:

Period Rate = Annual Interest Rate / Periods per Year

*Number of Periods = Periods per Year * Number of Years*

*= Annuity Payment *((1+ Period Rate)^No of Periods - 1/(Period Rate *(1 + Period Rate)^No of Periods*

Exercise 9

Find the present value of a semiannual ordinary annuity of $3,000 for seven years at 12% annual interest, compounded semiannually.

1. Begin in a new area of the ANNUITY worksheet.
2. Enter the title and subtitle and row labels **Present Value, Ordinary Annuity,** *blank row,* **Annuity Payment,** *blank row,* **Annual Interest Rate,** *blank row,* **No Period Yr,** *blank row,* **Number of Years,** *blank row,* **Present Value.**
3. Enter the given facts.
4. Enter the Present Value formula. In our example it is =AC254*((1+(AC256/AC258))^(AC258*AC260)-1)/((AC256/AC258)*(1+(AC256/AC258))^(AC258*AC260)).

5. Select and name the range OAPRESENTVAL.
6. Format the range as shown in Figure 8-5 and save the spreadsheet.

Your spreadsheet should look like Figure 8-5.

AB251	Present Value		
	AB	**AC**	**AD**
251	**Present Value**		
252	Ordinary Annuity		
253			
254	**Annuity Payment**	$ 3,000.00	
255			
256	**Annual Interest Rate**	12.0%	
257			
258	**No Period Yr**	2	
259			
260	**Number of Years**	7	
261			
262	**Present Value**	$27,884.95	
263			
264			
265			
266			
267			
268	**Effective Rate**	12.4%	
269			

Figure 8-5: The present value of an Ordinary Annuity shows how much a lump-sum savings plan would produce in the same time at the same rate.

Exercise 10

Stein and Company has established a sinking fund to retire a bond issue of $500,000, which is due in two years. How much is the quarterly sinking fund payment if the account pays 8% compounded quarterly? What lump sum amount could they deposit now in a similar compound interest savings plan to have the same return? Write a memo explaining their options.

Applicaton Exercises

Prepare worksheets for each of the following.

1. Len and Sharon Smith are saving money for their daughter, Heather, to attend college. They set aside an ordinary annuity of $4,000 annually at 7% annual interest. How much will Heather have for college when she graduates from high school in three years? Show the details of the plan.

2. Harry Taylor has a plan to pay an ordinary annuity of $5,000 annually for three years so that he can take a year away from work to study for a Master's degree in business. The annual rate of interest is 8%. How much will Harry have at the end of three years?

3. Scott Martin is planning to establish a small business to provide consulting services in computer networking. He is committed to an ordinary annuity of $3,000 annually at 8.5% annual interest. How much will Scott have to establish the business after three years?

4. Latanya Brown established an ordinary annuity of $1,000 annually at 7% annual interest. What is the future value of the annuity after 15 years? How much will Latanya have invested of her own money during this period of time? By how much will her investment have grown?

5. You invest in an ordinary annuity of $2,000 annually at 8% annual interest. What is the future value of the annuity at the end of five years? How much have you invested? How much interest has your annuity earned?

6. Pat Lechleiter pays an ordinary annuity of $2,500 quarterly at 8% annual interest compounded quarterly, to establish supplemental income for retirement. How much will Pat have available at the end of five years?

7. You invest in an ordinary annuity of $4,500 annually at 8% annual interest. Find the future value of the annuity at the end of ten years. How much interest has your annuity earned?

8. Bernard McGhee has decided to establish an annuity due of $2,500 annually for 15 years at 7.2% annual interest. How much is the annuity worth after two years?

9. Which annuity earns more interest: an annuity due of $300 quarterly for one year at 8% annual interest compounded quarterly, or an annuity due of $600 semiannually for one year at 8% annual interest compounded semiannually?

10. You have carefully examined your budget and determined that you can only manage to set aside $250 per year. So you set up an annuity due of $250 annually at 7% annual interest. How much of your own money will you have contributed after 20 years? What is the future value of your annuity after 20 years? How much interest will your money earn for you?

11. The Smiths have four children to send to school. Find the future value of an ordinary monthly annuity of $50 for 7 years at 5% annual interest compounded quarterly. How much will each child receive?

12. Ben and Susan Duke have a one-year-old daughter and are establishing a college fund. They want to save enough now to have her entire college expenses paid for at the time she enters college (17 years form now). If her college expenses are projected to be $35,000 for a two-year degree, what annual sinking fund payment should they make if the annual interest is 8%?

13. Bertha Looney recognizes the value of saving part of her income. She has set a goal to have $25,000 in cash available for emergencies. How much should she invest semiannually to have $25,000 in ten years if the sinking fund she has selected pays 8% annually, compounded semiannually?

14. Michelle and Joe Hanover have a 12-year-old daughter and are now in a financial position to begin saving for her college education. What annual sinking fund payment will they need to make to have her entire college expenses available at the time she enters college six years form now? Her college expenses are projected to be $30,000, and the annual interest rate is 6%.

15. Cindy Meziere plans to set aside $2,500 at the end of each year for two years at 8% annual interest to pay for her college tuition and expenses at a local community college. What lump sum could she invest today to achieve the same value at the end of two years?

16. Laura Kleinaitis is setting up an annuity for a memorial scholarship. She can set aside $3,000 at the end of each year for the next ten years and it will earn 7% annual interest. What lump sum would she need to set aside today at 7% annual interest to have the same scholarship fund available ten years form now?

17. George and Kathy Miller have agreed to pay for their granddaughter's college education. Melanie will be ready to attend college seven years from now. George and Kathy are trying to decide whether to deposit $2,300 at the end of each year for seven years into an annuity at 7% annual interest, or make a lump sum deposit now at 7% annual interest. What lump sum should be deposited now to have the same amount as the annuity after seven years?

18. William Farris, a nationally recognized philanthropist, set up an annuity by depositing $1,600 at the end of each year for ten years at 9% annual interest. The Farris fund would be used to reward faculty for teaching, service, and research at a major two-year community/technical college. How much would Farris have to deposit today at an annual interest rate of 9% to have the same amount as the annuity yields after the ten years?

19. Janice and Terry Van Dyke are investigating their retirement. Since they have ten more years until retirement, they have decided to establish a quarterly ordinary annuity of $3,000 for the next ten years at 8% annual interest compounded quarterly. How much should they invest in a lump sum now at 8% annual interest compounded quarterly if they want to have the same amount as the annuity yields after ten years?

20. The Shari Joy Corporation decided to set aside $3,200 at the beginning of every six months to provide donation funds for a new Little League baseball field scheduled to be built in 18 months. If money earns 12% annual interest compounded semiannually, how much will be available as a donation for the field?

21. A business deposits $4,500 at the end of each quarter in an account that earns 8% annual interest compounded quarterly. What is the value of the annuity in five years?

22. University Trailers is setting aside $800 at the beginning of every quarter to purchase a forklift in 30 months. The annual interest will be 8%, compounded quarterly. How much will be available for the purchase?

23. Lausanne Private School System needs to set aside funds for a new mainframe computer. What monthly sinking fund payment would be required to amount to $45,000, the approximate cost of the computer, in 1.5 years at 12% annual interest, compounded monthly?

24. Goldie's Department Store has a fleet of delivery trucks that will last for three years of heavy use and then need to be replaced at a cost of $75,000. How much must they set aside every three months in a sinking fund at 8% annual interest, compounded quarterly, to have enough money to replace the trucks?

25. Tasty Food Manufacturers, Inc., has a bond issue of $1,400,000 due in 30 years. If they want to establish a sinking fund to meet this obligation, how much must be set aside at the end of each year if the annual interest rate is 6%?

26. Danny Lawrence Properties, Inc., has a bond issue that will mature in 25 years for $1,000,000. How much must the company set aside each year in a sinking fund at 12% annual interest to meet this future obligation? What lump sum investment would be equal to this?

27. Mike's Sport Shop deposited $3,400 at the end of each year for 12 years at 7% annual interest. How much will Mike have in the account at the end of the time period? What lump sum saving deposit would be equal to this?

28. How much must be set aside at the end of each year by the Caroline Cab Company to replace four taxicabs at a cost of $90,000? The current interest rate is 6% annually. The existing cabs will wear out in three years.

29. Edward Spencer owns a lawn maintenance business. His riding lawnmower cost $850 and should last for six years. Edward wants to establish a sinking fund to buy a new mower. How much must he set aside at the end of each year at 12% annual interest to have enough money to buy the new equipment? If he can get a 7.7% rate on a quarterly compounded regular savings plan, what lump sum would he need to invest at this time?

30. How much will you need to invest today to have the same amount after 5 years as a quarterly ordinary annuity of $800 if both interest rates are 8% annually, compounded quarterly?

9. Mortgages

Mortgages are investments in a house or other building and property. They are also debts because when you or your business purchases a building or land, you agree to pay off a loan. On the other hand, buying a home or building builds equity and the value of the property, if all goes well, will increase over the years. The property is used as the collateral for the loan. In this chapter you will learn how to set up a spreadsheet for computing the monthly payment of a fixed-rate mortgage and create an amortization schedule template.

Objectives

By working through and completing the exercises below, you will be able to:

- Create and explain a Monthly Payment per $1,000 of Amount Financed table.
- Use the MATCH and INDEX functions to use values from a table.
- Create a template for and find the total interest and monthly payment for a fixed rate mortgage.
- Create a template for and use an amortization schedule to show the details of paying a fixed rate mortgage.

Monthly Payments

Mortgages involve regular payments of equal amounts over time, so they are annuities. The rate of interest on a loan for a fixed-rate mortgage remains the same for the entire time of the loan. Investors have several payment options. The number of years may vary. 15- and 30-year mortgages are the most common. When you have a fixed-rate mortgage the payment remains the same per month for the entire life of the loan.

Monthly Payment per $1,000 of Amount Financed Lookup Table

To set up a spreadsheet for computing a fixed mortgage payment you will first create a table. This is the Monthly Payment per $1,000 of Amount Financed table. With it you will use the INDEX function to find a value in a table. The values in this table are used in a formula to compute the monthly mortgage payment.

Exercise 1

Create the Monthly Payment per $1,000 of Amount Financed table.

1. Open a new spreadsheet (or new worksheet) and name it MORTGAGE.
2. Go to cell AA98.
3. Create the title of the table: **Monthly Payment per $1,000 of Amount Financed**
4. Starting in cell AB100 enter a series of decimal values across the row from .08 to .16 in increments of .05. The last value (.16) should finish in cell AR100. These represent annual interest rates.
5. Format this range of cells as percents.
6. Starting in cell AA101 and entering values down, enter a series of whole numbers beginning with 10 and incrementing by 1 up to 35 down the column. You should finish at row 126. These represent the number of years.

7. In cell AB101 enter a PMT function to find the payment for a $1,000 loan using a partially absolute reference to the column of years and a partially absolute reference to the interest rate column. Enter the formula: **=PMT(AB$100/12,$AA101*12,-1000)**.
8. Copy cell AB101 and select the range from AB101 to AR126.
9. Paste the formula into the range. The entire range should reflect the payment for a $1,000 loan for each corresponding year and interest rate (see Figure 9-1).
10. Select the range from AA100 to AR126 and name it FACTOR.
11. Select the range from AA100 to AR100 and name this row RATE.
12. Select the range from AA100 to AA126 and name this column TERM.
13. Format the range as shown in Figure 9-1 and save the spreadsheet.

AB102 =PMT(AB$101/12,$AA102*12,-1000)

Monthly Payment per $1,000 of Amount Financed

1		8.0%	8.5%	9.0%	9.5%	10.0%	10.5%	11.0%	11.5%	12.0%	12.5%	13.0%	13.5%	14.0%	14.5%	15.0%	15.5%	16.0%
2	10	12.13	12.40	12.67	12.94	13.22	13.49	13.78	14.06	14.35	14.64	14.93	15.23	15.53	15.83	16.13	16.44	16.75
3	11	11.42	11.69	11.96	12.24	12.52	12.80	13.09	13.38	13.68	13.98	14.28	14.58	14.89	15.20	15.51	15.82	16.14
4	12	10.82	11.10	11.38	11.66	11.95	12.24	12.54	12.83	13.13	13.44	13.75	14.06	14.37	14.69	15.01	15.33	15.66
5	13	10.33	10.61	10.90	11.19	11.48	11.78	12.08	12.38	12.69	13.00	13.31	13.63	13.95	14.28	14.60	14.93	15.27
6	14	9.91	10.20	10.49	10.78	11.08	11.38	11.69	12.00	12.31	12.63	12.95	13.28	13.60	13.94	14.27	14.61	14.95
7	15	9.56	9.85	10.14	10.44	10.75	11.05	11.37	11.68	12.00	12.33	12.65	12.98	13.32	13.66	14.00	14.34	14.69
8	16	9.25	9.54	9.85	10.15	10.46	10.77	11.09	11.41	11.74	12.07	12.40	12.74	13.08	13.42	13.77	14.12	14.47
9	17	8.98	9.28	9.59	9.90	10.21	10.53	10.85	11.18	11.51	11.85	12.19	12.53	12.87	13.22	13.58	13.93	14.29
10	18	8.75	9.05	9.36	9.68	10.00	10.32	10.65	10.98	11.32	11.66	12.00	12.35	12.70	13.06	13.42	13.78	14.14
11	19	8.55	8.85	9.17	9.49	9.81	10.14	10.47	10.81	11.15	11.50	11.85	12.20	12.56	12.92	13.28	13.65	14.02
12	20	8.36	8.68	9.00	9.32	9.65	9.98	10.32	10.66	11.01	11.36	11.72	12.07	12.44	12.80	13.17	13.54	13.91
13	21	8.20	8.52	8.85	9.17	9.51	9.85	10.19	10.54	10.89	11.24	11.60	11.96	12.33	12.70	13.07	13.45	13.82
14	22	8.06	8.38	8.71	9.04	9.38	9.73	10.07	10.42	10.78	11.14	11.50	11.87	12.24	12.61	12.99	13.37	13.75
15	23	7.93	8.26	8.59	8.93	9.27	9.62	9.97	10.33	10.69	11.05	11.42	11.79	12.16	12.54	12.92	13.30	13.69
16	24	7.82	8.15	8.49	8.83	9.17	9.52	9.88	10.24	10.60	10.97	11.34	11.72	12.10	12.48	12.86	13.25	13.63
17	25	7.72	8.05	8.39	8.74	9.09	9.44	9.80	10.16	10.53	10.90	11.28	11.66	12.04	12.42	12.81	13.20	13.59
18	26	7.63	7.96	8.31	8.66	9.01	9.37	9.73	10.10	10.47	10.84	11.22	11.60	11.99	12.38	12.76	13.16	13.55
19	27	7.54	7.88	8.23	8.58	8.94	9.30	9.67	10.04	10.41	10.79	11.17	11.56	11.95	12.34	12.73	13.12	13.52
20	28	7.47	7.81	8.16	8.52	8.88	9.25	9.61	9.99	10.37	10.75	11.13	11.52	11.91	12.30	12.70	13.09	13.49
21	29	7.40	7.75	8.10	8.46	8.82	9.19	9.57	9.94	10.32	10.71	11.09	11.48	11.88	12.27	12.67	13.07	13.47
22	30	7.34	7.69	8.05	8.41	8.78	9.15	9.52	9.90	10.29	10.67	11.06	11.45	11.85	12.25	12.64	13.05	13.45
23	31	7.28	7.64	8.00	8.36	8.73	9.11	9.48	9.87	10.25	10.64	11.03	11.43	11.82	12.22	12.62	13.03	13.43
24	32	7.23	7.59	7.95	8.32	8.69	9.07	9.45	9.84	10.22	10.62	11.01	11.41	11.80	12.20	12.61	13.01	13.42
25	33	7.18	7.54	7.91	8.28	8.66	9.04	9.42	9.81	10.20	10.59	10.99	11.39	11.79	12.19	12.59	13.00	13.40
26	34	7.14	7.50	7.87	8.25	8.63	9.01	9.39	9.78	10.18	10.57	10.97	11.37	11.77	12.17	12.58	12.99	13.39
27	35	7.10	7.47	7.84	8.22	8.60	8.98	9.37	9.76	10.16	10.55	10.95	11.35	11.76	12.16	12.57	12.98	13.38

Figure 9-1: Monthly Payment per $1,000 of Amount Financed.

Total Interest and Monthly Payments

The Monthly Payment per $1,000 is a handy reference tool for real estate salespeople and banks to quickly give payment estimates to clients. It can also be useful to a spreadsheet that calculates mortgage payments. However, it provides the same information as the PMT function you used in Chapter 8. For the sake of illustration, since some day you will have a need for lookup and reference functions, we will use this table.

The MATCH function is a lookup and reference function and returns the number of the row or column in which it finds the value you designate in a range of cells along a row or down a column.

To find a row number =MATCH(value, range, type)

where the value is the amount or text you are looking for, the range is the array of cells in a row or column, and the type is -1, 0, or 1. -1 finds the smallest value equal to or greater than the given value. Zero finds only a perfect match, and 1 finds the largest value equal to or less than the given value.

Once you know or have a reference to a row number and column number, you can use the INDEX function to get the value of a cell for that row and in any column you name.

To find a cell's value: =INDEX(range,column number,row number)

Using these two functions will return the monthly payment per $1,000 so by multiplying that value times the amount financed divided by $1,000 will show the monthly payment. The formula is

=INDEX(FACTOR,MATCH(Term in years,TERM),MATCH(Annual Interest Rate,RATE)) (Amount Financed/1000)*

Exercise 2

Lunelle Miller is purchasing a home for $87,000. Home Federal Savings and Loan has approved her loan application for a 30-year fixed-rate loan at 10% annual interest. If Lunelle agrees to pay 20% of the purchase price as a down payment, calculate the monthly payment. Other costs include the annual insurance premium at $923 and the annual tax on the property at $950.

1. Open the MORTGAGE spreadsheet.
2. Enter the title and labels: **Mortgage Payment**, *blank row*, **Purchase Price**, *blank row*, **Down Payment Rate, Down Payment**, *blank row*, **Amount Financed**, *blank row*,, **Annual Interest Rate, Term in years**, *blank row*, **Monthly Payment (lookup formula), Monthly Payment Function**, *blank row*, **Total Interest, Total Cost**, *blank row*, **Other Costs, Annual Insurance, Annual Property Tax**, *blank row*, **Total Other Costs**, *blank row*, **Monthly Payment for Taxes and Insurance**, *blank row*, **Adjusted Monthly Payment**
3. Enter the given data from the problem statement putting the currency amounts two columns from the labels and the rates and intermittent figures in the first column from the labels. In our example the Purchase Price is in column C and the Down Payment Rate is in column B (see Figure 9-2).
4. Name the Annual Interest value the ANNUALINTEREST range.
5. Enter the formula to compute the down payment: Purchase Price * Down payment rate, which in our example is =C3*B5.
6. Subtract the Down Payment from the Purchase Price to find the Amount Financed: =C3-C6.
7. Name the Amount Financed value's cell, the AMTFINANCED range.
8. Enter the monthly payment formula. In our example it is =INDEX(FACTOR,MATCH(B12,TERM),MATCH(B11,RATE))*(C8/1000)
9. Name the Monthly Payment value's cell the MONTHLYPAYMENT range.
10. Optionally, enter the formula for the Monthly Payment Function. In our example it is =PMT(B11/12,B12*12,-C8).
11. Enter a formula to find the total cost: =(MONTHLYPAYMENT *(Term in Years*12))+Down Payment, which in our example is =(MONTHLYPAYMENT *(B12*12))+C6.
12. Enter a formula to find the Total Interest: =Total Cost - AMTFINANCED, which, in our example is =C18- AMTFINANCED.
13. Total the Other costs including the extra row for future expansion. In our example the formula is =SUM(B21:B23).
14. Compute the monthly payment for other costs: =Total Other Cost/12, which in our example is =B24/12.
15. Add the Monthly Payment and the Monthly Payment for Other Costs to find the Adjusted Monthly Payment. In our example the formula is =C15+C26.
16. Select the range of used cells and name it MORTPAYMENT.

17. Format the spreadsheet as shown in Figure 9-2 and save the spreadsheet.

A1 | Mortgage Payment

	A	B	C	D
1	**Mortgage Payment**			
2				
3	**Purchase Price**		$ 87,000.00	
4				
5	**Down Payment Rate**	20%		
6	**Down Payment**		$ 17,400.00	
7				
8	**Amount Financed**		$ 69,600.00	
9				
10				
11	**Annual Interest Rate**	10%		
12	**Term in years**	30		
13				
14	**Monthly Payment (lookup formula)**		$ 610.79	
15	**Monthly Payment Function**		$ 610.79	
16				
17	**Total Interest**		$167,684.33	
18	**Total Cost**		$237,284.33	
19				
20	**Other Costs**			
21	**Annual Insurance**	$ 923.00		
22	**Annual Property Tax**	$ 950.00		
23				
24	**Total Other Costs**	$ 1,873.00		
25				
26	**Monthly Payment for Other Costs**		$ 156.08	
27				
28	**Adjusted Monthly Payment**		$766.87	
29				

Figure 9-2: With this portion of your spreadsheet you use the MATCH and INDEX functions to find the monthly payment.

The Amortization Schedule

An amortization schedule comes with large personal bank loans in the form of a monthly payment book. The client can pay each month and keep track of how much has been paid to date and how much is still due. The spreadsheet version is another running balance example.

Exercise 3

Create an amortization schedule for Lunelle's mortgage.

1. Open the MORTGAGE spreadsheet.
2. To the right and below the MORTPAYMENT range, enter the title, **Amortization Schedule**.
3. Skip a row and enter the column headings: **Month, Payment to Interest, Payment to Principal, End-of-Month Principal**.
4. Starting in the first cell in the Month column, create a series of numbers from 1 to 359 incrementing by one.
5. The First Payment to Interest formula is =AMTFINANCED*(ANNUALINTEREST/12).
6. The Payment to Principal formula is =MONTHLYPAYMENT-H40.
7. The first End-of-Month Principal formula is =AMTFINANCED-I40.
8. The second and subsequent Payments to Interest formula is =J40*(ANNUALINTEREST/12).

9. The second and subsequent End-of-Month Principal formula is the running balance formula: =J40-I41.
10. Copy the subsequent Interest and End-of-Month Principal and the Payment to Principal formulas down to payment 359.
11. Select this range of cells plus an extra row for expansion and name it AMORTIZE.
12. Format the range as in Figure 9-3 and save the spreadsheet.

Your amortization schedule will now show the complete process for repaying the mortgage loan (see Figure 9-3).

G38 | Amortization Schedule

Amortization Schedule

Month	Payment to Interest	Payment to Pincipal	End-of-Month Principal
1	$ 580.00	$ 30.79	$ 69,569.21
2	$ 579.74	$ 31.05	$ 69,538.16
3	$ 579.48	$ 31.31	$ 69,506.86
4	$ 579.22	$ 31.57	$ 69,475.29
5	$ 578.96	$ 31.83	$ 69,443.46
6	$ 578.70	$ 32.09	$ 69,411.37
7	$ 578.43	$ 32.36	$ 69,379.01
8	$ 578.16	$ 32.63	$ 69,346.38
9	$ 577.89	$ 32.90	$ 69,313.47
10	$ 577.61	$ 33.18	$ 69,280.30
11	$ 577.34	$ 33.45	$ 69,246.84
12	$ 577.06	$ 33.73	$ 69,213.11
13	$ 576.78	$ 34.01	$ 69,179.09
14	$ 576.49	$ 34.30	$ 69,144.80
15	$ 576.21	$ 34.58	$ 69,110.21
16	$ 575.92	$ 34.87	$ 69,075.34
17	$ 575.63	$ 35.16	$ 69,040.18
18	$ 575.33	$ 35.45	$ 69,004.73
19	$ [illegible]	[illegible]5.75	$ [illegible]
[illegible]	[illegible]	[illegible]	[illegible]
[illegible]	[illegible].61	$ [illegible]	[illegible]7.41
[illegible]	130.65	$ [illegible]	15,197.26
333	$ 126.64	$ 484.15	$ 14,713.12
334	$ 122.61	$ 488.18	$ 14,224.94
335	$ 118.54	$ 492.25	$ 13,732.69
336	$ 114.44	$ 496.35	$ 13,236.34
337	$ 110.30	$ 500.49	$ 12,735.85
338	$ 106.13	$ 504.66	$ 12,231.19
339	$ 101.93	$ 508.86	$ 11,722.33
340	$ 97.69	$ 513.10	$ 11,209.23
341	$ 93.41	$ 517.38	$ 10,691.85
342	$ 89.10	$ 521.69	$ 10,170.16
343	$ 84.75	$ 526.04	$ 9,644.12
344	$ 80.37	$ 530.42	$ 9,113.69
345	$ 75.95	$ 534.84	$ 8,578.85
346	$ 71.49	$ 539.30	$ 8,039.55
347	$ 67.00	$ 543.79	$ 7,495.76
348	$ 62.46	$ 548.33	$ 6,947.43
349	$ 57.90	$ 552.89	$ 6,394.54
350	$ 53.29	$ 557.50	$ 5,837.04
351	$ 48.64	$ 562.15	$ 5,274.89
352	$ 43.96	$ 566.83	$ 4,708.06
353	$ 39.23	$ 571.56	$ 4,136.50
354	$ 34.47	$ 576.32	$ 3,560.18
355	$ 29.67	$ 581.12	$ 2,979.06
356	$ 24.83	$ 585.96	$ 2,393.10
357	$ 19.94	$ 590.85	$ 1,802.25
358	$ 15.02	$ 595.77	$ 1,206.48
359	$ 10.05	$ 600.74	$ 605.74

Figure 9-3: The Amortization Schedule provides a detailed picture of a property investment.

Application Exercises

Ignoring other costs, prepare a report to explain the mortgage details for the following customers:

1. Phylis Wright is buying a home that sells for $67,000. The down payment must be 15% of the selling price. It is a 20-year mortgage at 8.5% interest.

2. Add comparative figures for Exercise 1 if it is financed for 25 years at 8.5%.

3. Stephen Black has just purchased a home for $155,000. Northridge Mortgage Company has approved his loan application for a 25-year fixed-rate loan at 9.5%. Stephen has agreed to pay 18% of the purchase price as a down payment.

4. Explain to Stephen Black his situation if he could budget an additional $100 for his monthly payment.

5. Justin Wimmer will pay $69.700 for a home at 8.5% interest with a 20-year fixed-rate loan and 10% down.

6. Heike Drechsler is will purchase a $84,700 home in the mountains. The 17-year fixed-rate loan has an interest rate of 9%. He must make an 18% down payment.

7. Conchita Martinez has agreed to pay $210,300 for a home near Albany, New York. Her 20-year fixed-rate loan has an interest rate of 8.5%. She will pay 20% down.

8. Hullett Houpt is purchasing a home for $97,000. He will finance the mortgage for 15 years and pay 11% interest on the loan. He makes a down payment that is 20% of the purchase price.

9. Bob and Janice Malena need to finance $80,000 on their new home. After checking with several mortgage companies they have narrowed their choices to two options: 20 years at 11% and 25 years at 10%. Explain their options to them and make a recommendation.

10. Ginger Canoy has purchased a home for $122,000. She must make an 18.5% down payment and pay a mortgage for 15 years at 9.5% interest.

11. Qua Wau is trying to determine whether to accept a 25-year 9% mortgage or a 20-year 9% mortgage on the house he is planning to buy. He needs to finance $125,700 and has planned to budget $1,250 monthly for his payment. Prepare a written report advising him.

12. The Townsends purchased a $98,000 home by making an $8,000 down payment and signing a 10% mortgage with monthly payments for 25 years.

13. Mr. and Mrs. Rivera qualified for an $85,000 home by making a $5,000 down payment. The current mortgage rate is 9.75%. Explain how different the loan would look if it were for 20 versus 25 years.

14. Mr. and Mrs. Blood purchased a $320,000 home by making a $20,000 down payment at 9.25% interest. Explain their best decision if offered a 25- or 30-year mortgage.

15. Find a building in your town or city you would like to own for business or home. Write a report showing how you might finance the purchase of this property.

10. Stocks and Bonds

Working with stocks and bonds will give you another opportunity to build and use tabled data. Both stock and bond information is available in your daily paper and you can enter this information into your Stocks and Bonds spreadsheet as needed. Now days, of course, the way to make this whole process very automatic is to have access to the data on-line through a service possibly on the Internet. This chapter will not give you the experience and knowledge needed to "play the stock market," but it will familiarize you with the mathematics involved in the game.

Objectives

By working through and completing the exercises below, you will be able to:

- Create and explain a Stock Listings table.
- Make a Disbursement of Dividends template and find the preferred and common stock dividends.
- Create a Cost of Buying template and find the price paid for stocks.
- Create a Proceeds from Selling template and find the proceeds from selling stock.
- Create a Return on Investment template and find the percent of loss or gain.
- Create and explain a Bonds Listing table.
- Create a Cost and ROI bonds template and find the return on investment for bonds..

Stocks

When working with stocks you must first be able to interpret the stock listings. You can find them in most daily newspapers and for the most complete report you will want to review the *Wall Street Journal* or other trade publications. In addition to keeping a table listing, you will want to keep tabs on the progress of your investment and understand how stocks are bought and sold. Part of your income from stocks comes from periodic (quarterly or yearly) dividends. Companies need to accurately compute the distribution of dividends when they declare them.

Stock Listing

In Exercise 1 you will create a version of a stock listing. You will be able to use this information to make decisions on when to buy, sell, or hold on to a stock. You will include five formulas to help interpret the meaning of the figures. First, you will include the Annual Earnings per share figure to compute the Price to Earnings Ratio. You will also compute the Current Stock Yield figure. Finally, you will find the previous day's price of the stock and the current cost of the stock. All four of these calculations help the experienced trader make sound decisions about their holdings.

Exercise 1

Begin a stock listing table.

1. Open a new spreadsheet and save it as STOCKS or start a new worksheet in the INVEST spreadsheet.
2. Enter the title, **My Stocks**.
3. Beginning in cell A3 create a series from 1 to 18 across the row.

4. In cell B4, enter 52 Weeks and center it across columns B and C.
5. Starting in Cell A5, enter the labels across the row: **No., High, Low, Stock, Dividend, Yield %, Annual Earnings pr Share, PE Ratio, Sales 100s, High, Low, High - Low, Last, Change, Previous Day's Closing, Price Bought At, Number Shares Own, Current Value.**
6. Enter data for Avnet and ATMOS stocks from Table 10-1.

Table 10-1

High	30	23	6	51
Low	23 1/4	16 1/4	3 1/8	37 1/8
Stock	Avnet	ATMOS	Avalon	Avon
Dividend	0.60	1.24		1.40
Annual Earnings pr Share	1.47	1.31		1.81
Sales 100s	4524	521	84.00	9405
High	26 1/2	20 7/8	3 3/8	49
Low	25 1/4	20 1/4	3 1/8	47
Last	26 3/8	20 3/8	3 1/8	47
Change	7/8	-1/8	- 3/8	7/8
Price bought at	33 7/8	15 1/4	5 1/8	47
No shares own	100	125	368	368

7. Create a series incrementing by one down column A from A6 to A25 to see the numbers 1 through 20.
8. Select cells A5 to O5 and name it the MYSTOCKS range.
9. Enter the formulas from Table 10-2 to fill in the remaining cells:

Table 10-2

Name	*Formula*	Example
Current Stock Yield:	*IF(Dividend is greater than zero, then Dividend/Last, otherwise enter "...")*	=IF(E6>0,E6/M6, "...")
Price to Earnings Ratio (PE Ratio)	*=IF(Annual Earnings per Share is greater than zero, then Last/Annual Earnings per Share, otherwise "...")*	=IF(G6>0,M6/G6,"...").
High - Low selling prices (a crude measure of variance and might mean a stock is volatile and unsteady)	*High - Low*	=J6-K6.
Previous day's closing price (a wide difference could mean big profit or losses but also an unsteady investment)	*Last - Change*	=M6-N6..
Current value (how much your investment would be worth if sold immediately)	*=Last price * No of Shares Owned*	*=Q6*M6*

Note: *Annual Earnings per Year is not usually a part of the Stock Listings but the PE Ratio is.*

10. Copy the formulas down to row 25.
11. Select the range A3 to R26 and give it the MYSTOCKS name.

12. Format the table as you can see in Figure 10-1 and save the spreadsheet.

A1 | My Stocks

My Stocks

52 Weeks

No.	High	Low	Stock	Dividend	Yield %	Annual Earnings pr Share	PE Ratio	Sales 100s	High	Low	High - Low	Last	Change	Previous Day's Closing	Price bought at	No shares own	Current Value
1	30	23 1/4	Avnet	3/5	2.3%	$ 1.47	18	4,524.00	26 1/2	25 1/4	1 1/4	26 3/8	7/8	25 1/2	33 7/8	100	2,637.50
2					...		...							0			-
3					...		...							0			-
4					...		...							0			-
5					...		...							0			-
6					...		...							0			-
7					...		...							0			-
8					...		...							0			-
9					...		...							0			-
10					...		...							0			-
11					...		...							0			-
12					...		...							0			-
13					...		...							0			-
14					...		...							0			-
15					...		...							0			-
16					...		...							0			-
17					...		...							0			-
18					...		...							0			-
19					...		...							0			-
20					...		...							0			-

Figure 10-1: With this listing table you can quickly review your stock investments. Having the data entered from an on-line service makes it automatic.

Exercise 2

Write a report summarizing the Avalon stock information. Do you think you would sell, buy more, or hold this stock?

Dividends

Each dividend period (quarterly or annually) a public corporation's board of directors may declare a dividend. This may mean they have made enough profits to return money to their stockholders. Often no dividends are offered because the company deems it necessary to reinvest the profits back into the business.

Your dividend depends on the kind of stock you bought. If you bought Preferred stock you must be paid a set rate every period. However, if you bought Common stock, you will be paid an equal portion of whatever dividend is available after the preferred stockholders are paid. If a dividend is not paid in a period, preferred stockholders are owed the payment and this payment in "arrears" must be paid to them before any other dividend is paid.

Exercise 3

The Avalon board of directors have declared a dividend this quarter. It has issued 20,000 shares of cumulative preferred stock that will earn dividends at $.60 per share, and 100,000 shares of common stock. Last quarter they paid no dividends. They now have $250,000 available for dividends. Create a dividend distribution template and show how Avalon will distribute the dividends.

1. Go to a new section of the STOCK spreadsheet.
2. Create the heading **Distribution of Dividends**.
3. Skip a row and enter the following labels down the column: **Total Dividend,** *blank row,* **Shares of Preferred Stock, Shares of Common Stock,** *blank row,* **Total Stock,** *blank row,* **Preferred Dividend per Share,** *blank row,* **Was Dividend Paid Last Period?,** *blank row,* **Dividend in Arrears,** *blank row,* **Available after Arrears,** *blank row,* **Preferred Dividend,** *blank row,* **Common Dividend,** *blank row,* **Common per Share**.

4. Enter the given values with the Total Dividend two columns to the right and N (no) for the Was Dividend Paid Last Period? Value.
5. Enter the Total Stock formula: =Shares of Preferred Stock + Shares of Common Stock, which in our example is =T106+T107.
6. Enter the Arrears formula: =IF(No dividend last period,then Perferred Stock*Rate, otherwise 0), which in our example is =IF(T113="N",T106*T111,0).
7. Enter the formula for dividend Available after Arrears: =Total Dividend - Arrears, which in our example is =U104-U115.
8. Enter the formula for the Preferred Dividend for this period: =Preferred Stock*Rate, which in our example is =T106+T107.
9. If the Preferred Dividend were to use up all the available dividend, then the Common Dividend would be zero. Enter an IF formula to control for this possibility and also provide for the total Common Dividend (=Total Dividend - Preferred Dividend). In our example this is =IF(U117-U119>0,U117-U119,0).
10. Divide the Common Dividend by the number of Shares of Common Stock to find the Common per Share.
11. Format the Distribution of Dividends area as you see in Figure 10-2 and name it the DISTDIVIDENDS range.
12. Add this Avalon Common Dividend to the MYSTOCKS range in its Dividend column.
13. Save the spreadsheet.

S102 | Distribution of Dividends

	S	T	U	V
102	**Distribution of Dividends**			
103				
104	**Total Dividend**		$ 250,000.00	
105				
106	**Shares of Preferred Stock**	20000		
107	**Shares of Common Stock**	100000		
108				
109	**Total Stock**	120000		
110				
111	**Preferred Dividend per Share**	$ 0.60		
112				
113	**Was Dividend Paid Last Year?**	N		
114				
115	**Dividends in Arrears**		$ 12,000.00	
116				
117	**Available after Arrears**		$ 238,000.00	
118				
119	**Preferred Dividend**		$ 12,000.00	
120				
121	**Common Dividend**		$ 226,000.00	
122				
123	**Common per Share**	$ 2.26		
124				
125				

Figure 10-2: Use the Distribution of Dividends template to find how much to pay each stockholder.

Cost of Buying

Most investors buy and sell through a middle person called a stock broker or discount broker. Stock brokers charge more commission and provide financial counseling and other services. Discount brokers charge less commission on sales and purchases but do not provide other services.

Stocks are bought in Round lots of 100 shares. Anything leftover after buying your round lots is called an Odd lot. Brokers charge extra for dealing in odd lots.

Exercise 4

You decided to purchase 250 shares of Avon Products' stock on October 14, 1994. Your broker charged 2% of the stock price for trading round lots and an additional 1% for odd-lot portions. You will pay the current closing price.

1. Open the STOCK worksheet.
2. In the MYSTOCKS range, fill in the data for Avon stock from Table 10-1 on page 104.
3. One or two columns to the right of the DISTDIVIDENDS range, enter the Cost of Buying title.
4. Skip two rows down and enter the following labels: **Date Purchased, Round Lot, Odd Lot, Total Shares,** *blank row,* **Cost per share,** *blank row,* **Cost of Round Lot, Cost of Odd Lot,** *blank row,* **Commission Rate on Round Lot, Commission Rate on Odd Lot,** *blank row,* **Commission Round Lot, Commission Odd Lot,** *blank row,* **Total Commission,** *blank row,* **Total Cost.**.
5. Enter the given data from the problem statement above.
6. Enter the Last Price as found in the MYSTOCKS range.
7. Enter the Cost of the Round Lot formula two columns over from its label: =Round Lot * Cost per Share, which in our example is **=Z130*Z134**.
8. Enter the Cost of the Odd Lot formula beneath the Round Lot formula: =Odd Lot * Cost per Share, which in our example is **=Z131*Z134**.
9. The formula for the Commission Rate Odd Lot is: =Commission Rate on Round Lot + 1%. You can enter a formula for this or just enter .03.
10. Compute the commission on each lot multiplying their rates times their costs. In our example this is **=Z139*AA136** and **=AA137*Z140**
11. Total the commission below the costs. In our example this is **=SUM(Z142:Z143)**.
12. For total cost, enter a formula that adds the costs and the total commission: Cost Round + Cost Odd + Total Commission, which in our example is **=SUM(AA136:AA145)**.
13. Name the Total Cost value the COST range.
14. Name the range of cells COSTSTOCK.
15. Format the range as seen in Figure 10-3 and save the spreadsheet.

Compare your results with Figure 10-3.

Y127 Cost of Buying

	Y	Z	AA	AB
127	**Cost of Buying**			
128				
129	**Date Purchased**	10/14/94		
130	**Round Lot**	200		
131	**Odd Lot**	50		
132	**Total Shares**	250		
133				
134	**Cost per share**	$ 47.00		
135				
136	**Cost of Round Lot**		$ 9,400.00	
137	**Cost of Odd Lot**		$ 2,350.00	
138				
139	**Commission Rate on Round Lot**	2%		
140	**Commission Rate on Odd Lot**	3%		
141				
142	**Commission Round Lot**	$ 188.00		
143	**Commission Odd Lot**	$ 70.50		
144				
145	**Total Commission**		$ 258.50	
146				
147	**Total Cost**		$ 12,008.50	
148				
149				

Figure 10-3: Find the cost of your stock purchase.

Proceeds from Selling

Stocks can be a valuable source of income if business is good and dividends are regular. Sometimes, however, it becomes necessary for personal or business reasons to sell your stock. Hopefully you will sell at a profit. Brokers also take commissions for performing the sale. You might decide to sell at a price lower than what you paid if you have already recovered your investment in dividends.

Exercise 5

A year has passed since you bought the AVON stock. Avon is selling for $54 a share. You wish to sell Avon stock to invest elsewhere. Call your broker and have her sell all 250 Avon shares. During the year you received a dividend of $1 per share. Find the proceeds of the sale if the commission rates remain the same.

1. Open the STOCK worksheet.
2. Go to and copy the COSTSTOCK range a few columns to its right.
3. Change its title to **Proceeds from Selling**.
4. Select its cells and name it the PROCEEDSSTOCK range.
5. Edit the Date Purchased cell to read **Date Sold**.
6. Edit the Avon data in the MYSTOCKS range to reflect the $1 per share dividend, a $56 high and a $49 low, and a Last selling price of $54,
7. Go to the PROCEEDSSTOCK range.
8. Edit the Total Cost label to **Total Proceeds**.

9. Edit the Total Proceeds formula to subtract the commissions: =Cost Round + Cost Odd - Total Commission, which in our example is =AE136+AE137-AE145.
10. Name the cell the PROCEEDS range.
11. Format the range as shown in Figure 10-4 and save the spreadsheet.

Your spreadsheet should look like Figure 10-4 with proceeds equal to $13,203.

Y127 | Cost of Buying

	AC	AD	AE	AF
127	**Proceeds from Selling**			
128				
129	**Date Sold**	2/15/96		
130	**Round Lot**	200		
131	**Odd Lot**	50		
132	**Total Shares**	250		
133				
134	**Cost per share**	$ 54.00		
135				
136	**Cost of Round Lot**		$ 10,800.00	
137	**Cost of Odd Lot**		$ 2,700.00	
138				
139	**Commission Rate on Round Lot**	2%		
140	**Commission Rate on Odd Lot**	3%		
141				
142	**Commission Round Lot**	$ 216.00		
143	**Commission Odd Lot**	$ 81.00		
144				
145	**Total Commission**		$ 297.00	
146				
147	**Total Proceeds**		$ 13,203.00	
148				
149				

Figure 10-4: The proceeds from a sale of Avon stock at $54 a share.

Return on Investment

The Return on Investment (ROI) is income expressed as a percent of the cost stock. It takes into account your earnings from dividends received. The exercise has you selling stock after one dividend period but in real practice you will want to set up some way to keep records over time and reference your total dividends for computing the ROI. What is most interesting is the ROI per Year figure. Using this figure you can more accurately assess the long-term value of one stock over another. With the number of years expressed in days, the formulas for ROI are:

Number of Years - (Date Sold - Date Purchased)/365

Total ROI Gain = Proceeds - Cost + Total Dividend

*Total Dividend = Dividend Rate * Number of Shares Owned*

Return on Investment = Total ROI Gain/Number of Years

Exercise 6

You bought the AVON stock on October 14, 1994, and sold it on February 15, 1996. What is the return on your investment in Avon Products Stock?

1. Open the STOCK worksheet and go to the PROCEEDSSTOCK range.

2. Skip a column and enter the **Return on Investment** title.
3. Skip a row and enter the row labels: **Date Purchased, Date Sold**, *blank row,* **Cost of Buying, Cost of Selling**, *blank row,* **Number of Years**, *blank row,* **Dividend Rate**, *blank row,* **Shares Owned, Total Dividend**, *blank row,* **Total ROI Gain**, *blank row,* **Return on Investment**, *blank row,* **ROI per Year.**
4. Add the Date Purchased to the top of the COSTSTOCK range and the Date Sold to the top of the PROCEEDSSTOCK range.
5. Reference the Date Purchased and Date Sold from the COSTSTOCK and PROCEEDSSTOCK ranges respectively. In our example this is =Z129 and =AD129.
6. Enter the Number of Years formula. In our example this is =(AH130-AH129)/365.
7. Enter the Dividend Rate for the AVON stock from the MYSTOCKS range.
8. Enter the shares owned and the Total Dividend formula. In our example this is =AH137*AH139.
9. Enter the Total Gain formula. In our example this is =PROCEEDS-COST+AI140.
10. Enter the ROI per Year formula. In our example this is =AH144/AH135.
11. Select the range of cells used and name it ROI.
12. Format the ROI range as shown in Figure 10-5 and save the spreadsheet.

Y127 | Cost of Buying

	AG	AH	AI	AJ
127	**Return on Investment**			
128				
129	**Date Purchased**	10/14/94		
130	**Date Sold**	2/15/96		
131				
132	**Cost of Buying**		$ 12,008.50	
133	**Cost of Selling**		$ 13,203.00	
134				
135	**Number of Years**	1.34		
136				
137	**Dividend Rate**	$ 1.00		
138				
139	**Shares Owned**	250		
140	**Total Dividend**		$ 250.00	
141				
142	**Total ROI Gain**		$ 1,444.50	
143				
144	**Return on Investment**	12.0%		
145				
146	**ROI per Year**	9.0%		
147				

Figure 10-5: The Return on Investment shows the percent of gain (or loss) on an investment.

Bonds

The face value of a bond is its maturity value. In fact, in the trading of bonds a bond may sell for a greater or lesser amount than its face value.

Corporate bonds are issued by a business to raise cash. The bond's face value is usually $1,000 with a fixed rate of interest. The interest makes bonds an attractive investment. Bonds are generally more secure than stock because they represent a loan rather than part ownership. With part ownership comes part risk. However, if a business fails, stockholders get the leftovers after the debtors (including the bond holders) get paid first.

Bond Listing

Bonds are listed in a similar manner to stocks but with fewer complications. The only formula to include in a regular listing is the current yield. The formula for current yield is

=annual interest per bond/current price per bond

You will also enter a formula to compute the current price, the previous percent of $1,000, and the previous price. Note too that volume still means the number traded this day, however it is not in thousands as are stocks. Here 10 is 10 and 720 is 720.

Exercise 7

Create a bond listing work area in your spreadsheet.

1. Begin a BONDS spreadsheet or worksheet in the INVEST spreadsheet.
2. Enter the title centered across Twelve columns: **My Bonds**
3. Enter the labels across the row: **Bond No, Bond, Interest Rate, Year of Maturity, Present Percent of $1,000, Current Price,** ***blank column,*** **Net Change, Previous Percent of $1,000, Previous Price, Current Yield.**
4. Fill 20 rows with a series of numbers in the Bond No column.
5. Fill in the given values using Table 10-3.
6. Select the Bond column entries (not the label) and name it BONDNAMES.

Table 10-3

Bond	Abrnd	AmStor	Apache
Interest Rate	8 1/2		7 1/2
Year of Maturity	1999	2001	2000
Volume	29	105	720
Present Percent of $1,000	102.5%	103.5%	99.25%
Net Change	5/8	-1 1/2	- 3/4

6. Enter the Current Price formula for the Abrnd bond: =Present Percent of $1,000*1,000, which in our example is **=AO202*1000.**
7. Enter the Previous Percent of $1,000 formula for the Abrnd bond: =Present Percent of $1,000 - Net Change, which in our example is **=AQ202-AT202.**
8. Find the Previous Price: =Previous Percent of $1,000*1,000, which in our example is **=AU202*1000.**
9. Enter the formula for the Current Yield Percent. In our example it is **=AN202/AR202.**

10. Select the range of cells used plus an extra row for expansion and name it the MYBONDS range.
11. Copy the formulas down to the 20th row.
12. Format the table as you can see in Figure 10-6 and save your spreadsheet.

AL197 | My Bonds

My Bonds										
Bond No	Bond	Interest Rate	Year of Maturity	Volume	Present Percent of $1,000	Current Price	Net Change	Previous Percent of $1,000	Previous Price	Current Yield
1	ABrnd	8.500%	1999	29	102.50%	$ 1,025.00	5/8	2/5	$ 400.00	0.008%
2	AmStor		2001	105	103.50%	$ 1,035.00	-1 1/2	2 1/2	$ 2,535.00	
3	Apache	7.500%	2000	720	99.25%	$ 992.50	- 3/4	1 3/4	$ 1,742.50	0.008%
4	AMAX	8.625%	1999	1	96.25%	$ 962.50	1/4	5/7	$ 712.50	0.009%
5	Arml	13.500%	1999	279	105.00%	$ 1,050.00	1/2	5/9	$ 550.00	0.013%
6						$ -		0	$ -	
7						$ -		0	$ -	
8						$ -		0	$ -	
9						$ -		0	$ -	
10						$ -		0	$ -	
11						$ -		0	$ -	
12						$ -		0	$ -	
13						$ -		0	$ -	
14						$ -		0	$ -	
15						$ -		0	$ -	
16						$ -		0	$ -	
17						$ -		0	$ -	
18						$ -		0	$ -	
19						$ -		0	$ -	
20						$ -		0	$ -	

Figure 10-6: The Bonds Listing table you have made computes the necessary information.

Cost of Buying Bonds

Buying and selling bonds is quite the same situation as buying and selling stocks. The closing price is a percent of the face value. Premium bonds cost more than their face value but represent more secure investments. The bonds of a company gone bankrupt are called junk bonds and are only worth a fraction of their face value.

The payoff of a bond can be realized by either selling it or keeping it until maturity. If you keep it until maturity you can calculate a Return on Investment figure. The ROI formula for bonds is:

Total Gain/Cost of Purchase

Exercise 8

It is March 10, 1996 and Dottie Todd wants to by $10,000 in bonds, which would mean 10 bonds. After significant research she has narrowed her choices to AMAX and Arml. She will pay $5 per bond in commissions. If she plans to keep the bonds until maturity, which would yield the highest ROI and best ROI per year?

1. Open the BONDS worksheet and go to the MYBONDS range.

2. Set up a new area to the right with the title **Cost and ROI of Bonds**.

3. Enter the labels down the first column: **Bond,** *blank row,* **Date Purchased, Date Redeemed or Sold, Number of Bonds Purchased, Price per bond, Bond Cost, Broker's Fee Rate, Broker's Fee,** *blank row,* **Cost of Purchase, Face Value, Number of Years, Interest Rate,** *blank row,* **Interest Earned,** *blank row,* **Total Gain,** *blank row,* **Rate on Investment,** *blank row,* **ROI per year.**

4. Add the AMAX and Arml data to the MYBONDS range using Table 10-4.

Table 10-4

Bond	AMAX	Arml
Interest Rate	8.625%	13.500%
Year of Maturity	1999	1999
Volume	1	279
Present Percent of $1,000	96.25%	105.00%
Net Change	1/4	1/2

5. Begin with the AMAX bonds and enter the bond names and the given data including an absolute reference to the bond's current prices column in the MYBONDS range.

6. Enter the given data for Number of Bonds Purchased, Broker's Fee Rate, and the formula for the Number of Years (3/10/1999-3/10/1996).

Table 10-5

Name	Formula	Example
Bond Cost	=Price per Bond * Number of Bonds Purchased	=AZ205*AZ206
Broker's Fee	=Broker's Fee Rate * Number of Bonds Purchased	=AZ205*AZ210
Cost of Purchase	=Cost of Bonds + Broker's Fee	=AZ208+AZ211
Face Value of the bonds (you must enter the selling price directly)	=Number of Bonds Purchased * 1,000	=AZ205*1000
Enter an absolute reference to the Interest Rate from the MYBONDS range =$AN$202		
Interest Earned	=OR(Face Value,Selling Price) * No Years * Interest Rate	=AZ215*AZ216*AZ217
Total Gain is	=Face Value + Interest Earned – Cost of Purchase	=(AZ211*1000)+AZ225-AZ219
Return on Investment (ROI)	=Total Gain/Cost of Purchase	=AZ221/AZ208
ROI! Per Year	=ROI/Number of Years	=AZ223/AZ216

7. Copy all the values and formulas to the right to get the information for Arml bonds but edit the two absolute references to match the Arml data.

8. Select the range of cells used and name them BONDSROI.

9. Format (see Figure 10-7) and save the spreadsheet.

Check your work against Figure 10-7.

AY198 | Cost and ROI of Bonds

	AY	AZ	BA	BB
198	**Cost and ROI of Bonds**			
199				
200	**Bond**	AMAX	Arml	
201				
202	**Date Purchased**	3/10/96	3/10/96	
203	**Date Redeemed or Sold**	3/10/99	3/10/99	
204				
205	**Number of Bonds Purchased**	5	5	
206	**Price per bond**	$ 962.50	$ 1,050.00	
207				
208	**Bond Cost**	$ 4,812.50	$ 5,250.00	
209				
210	**Broker's Fee per Bond**	$ 5.00	$ 5.00	
211	**Total Broker's Fee**	$ 25.00	$ 25.00	
212				
213	**Cost of Purchase**	$ 4,837.50	$ 5,275.00	
214				
215	**Face Value or Selling Price**	$ 5,000.00	$ 5,000.00	
216	**Number of Years**	3.00	3.00	
217	**Interest Rate**	8.625%	13.500%	
218				
219	**Interest Earned**	$ 1,293.75	$ 2,025.00	
220				
221	**Total Gain**	$ 1,481.25	$ 1,775.00	
222				
223	**Rate on Investment**	31%	34%	
224				
225	**ROI per year**	10%	11%	
226				

Figure 10-7: Compare the ROIs of bonds before buying.

Exercise 9

Make the INVEST spreadsheet or the individual spreadsheets from this section into templates.

Application Exercises

Add the following information from to your stock listings and use it to solve problems 1–5 below.

Table 10-6

High	78	49 7/8	6 3/8	25 1/2
Low	44 1/2	34 1/4	0 1/8	18 1/2
Stock	HewlPk	Hilton	HmeFd	Honda
Dividend	.98	1.20	.	.21
Annual Earnings pr Share	$3.48	$1.76		$1.14
Sales 100s	38254	5564	8845	363
High	76 1/4	45 7/8	1 1/2	23 1/8
Low	72 7/8	42 3/8	1 1/8	22 3/4
Last	73	43 7/8	1 1/4	22 7/8
Change	– 2 1/2	– 1/8	– 1/4	– 1/8
Price bought at	57 5/8	41 1/4	2 7/8	22 3/4
No shares own	50	100	500	200

1. What is theYield, PE Ratio, High-Low, Previous Day's Closing, and Current Valueof eachstock
2. Sell 50 shares of Hilton at the last price. What is the cost of selling if the broker charges 2% per round lot and another 1% for odd lots.
3. Sell 300 shares of HmeFd. What is the ROI of those 300 shares?
4. Purchase 150 shares of HewlPk at the last price. Your broker charges 4% per round lot and another 1.5% for each odd lot. What is your cost?
5. Purchase 200 more shares of Honda. What is your cost given the same broker's fee rate as in number 4?
6. Artcraft Sales, Inc., has 1,000 shares of preferred stock outstanding and 10,000 shares of common stock. Preferred stockholders are guaranteed $5 per share dividends. The board declared a $50,000 dividend.
7. Because of low sales during the next year of operations, the Artcraft Sales, Inc., declared a dividend of only $35,000. The corporation still has 1,000 shares of preferred ($5 per share dividend) and 10,000 common. Though no dividend was declared last year, what are the Preferred and Common Dividends this year?
8. No dividend was paid last year by McGill, Inc. A dividend of $325,000 has now been declared for the current year. Stockholders own 100,000 of common stock and 25,000 shares of ($3) cumulative preferred stock.
9. Good Times Video, Inc., declared a dividend of $160,000 this year, having skipped paying a dividend last year. There are 10,000 shares of common stock and 18,000 shares of 8%, $100 par-value ($8 per share) preferred stock. What is the common per share?
10. In order to increase its retained earnings, the board of Wyatt Distributors, Inc., has not declared a dividend during the previous 2 years. This year, however, a $366,000 dividend has been approved. They have 55,000 shares of common stock and 6,000 shares of ($7.50) preferred stock. What is the dividend in arrears?

Use the following Table 10-7 to complete problems 11 to 15 below.

Table 10-7

Bond	Revlln	Mobil	Mobil	AMAX
Interest Rate	11.75%	7.625%	8.625%	14.5%
Year of Maturity	1999	2001	2004	1999
Volume	91	168	235	257
Present Percent of $1,000	89.875%	94.875%	95.125%	115%
Net Change	3/8	-7/8	-5/8	– 1/4

11. What are the Current Price, Previous % of 1,000, Previous Price and Current Yield of each bond.

12. Assume that three years ago you purchased five Revlln bonds at 85% of face value and the commission per bond was $5. What will your ROI be at maturity?

13. Assume that you bought five Revlln bonds at 87.125% and you sell all five this day at 89.875%. Write a report showing your gain or loss.

14. What is the total purchase of 15 of each of the Mobil bonds at today's price? If it is now 1996, what would the investment be worth at maturity? What would the ROI be?

15. Purchase 20 AMAX14.5 bonds. If this is 1996, write a report explaining the value of the investment at maturity.

Challenge Problem: Using your local paper or the Wall Street Journal or even the Internet, gather the prices of a set of stocks and bonds for three weeks. Prepare daily notes on your analyses and report on the advisability of buying the different offerings.

IV. Buying and Selling

11. Trade Discounts

Manufacturers and distributors give trade discounts as incentives for a sales. In this chapter you will learn how to set up spreadsheet templates to quickly compute the value of a single trade discount and the net price for a trade discount series.

Objectives

By working through and completing the exercises below, you will be able to:

- Create and use a single trade discount template.
- Create and use a trade discount series template.

Single Trade Discount

Manufacturers describe their products in a catalog. Wholesalers or retailers will see the suggested retail price. This price is also known as the Catalog Price or the List Price. When you or your business buys direct from the manufacturer, you will not pay the list price because you get a trade discount.

The trade discount formula is a very straight forward percentage process answering the question, "What is the net price after discount?" The formulas used are:

$$Discount = List\ Price * Discount\ Rate$$

$$Net\ Price = List\ Price - Discount$$

The net price formula can become a single step process by using the complement of the discount rate:

$$Net\ Price = List\ Price * (1\text{-}Discount\ Rate)$$

Exercise 1

The list price of a refrigerator is $600. Young's Appliance Store can buy the refrigerator at the list price less 20%. What is the discount? What is the net price of the refrigerator?

1. Open a new spreadsheet and save it as TRADE or, if you will be using multiple worksheets for this Buying and Selling section, start a new worksheet called TRADE DISCOUNT in the new BUYSELL spreadsheet.
2. Enter the title **Single Trade Discount**.
3. Skip a row and create the labels: **List Price,** ***blank row,*** **Discount Rate, Trade Discount,** ***blank row,*** **Net Price**
4. In the first column next to the labels enter the Discount rate.
5. In the second column enter the List Price.
6. Enter the formula for discount. In our example this is **=C3*B5**.
7. Format a line beneath the trade discount.
8. Enter the formula for the Net Price. In our example this is =C3-C6 or =C3*(1-B5)
9. Select the range of cells used and give it the DISCOUNTSINGLE name.
10. Format the range as you can see in Figure 11-1.
11. Save the spreadsheet.

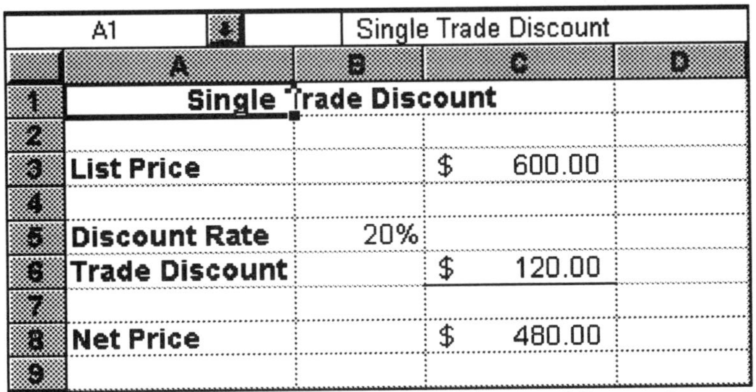

A1 | Single Trade Discount

	A	B	C	D
1	**Single Trade Discount**			
2				
3	**List Price**		$ 600.00	
4				
5	**Discount Rate**	20%		
6	**Trade Discount**		$ 120.00	
7				
8	**Net Price**		$ 480.00	
9				

Figure 11-1: The single trade discount is one of the easier math problems in business.

Exercise 2

Modify the Single Trade Discount to reflect the purchase of 20 refrigerators at the given discount rate. Simply add a row for the Number of items purchased and a formula that multiplies this times the list price.

Trade Discount Series

Manufacturers often provide more than one discount on the same purchase. They may do so to promote a product or increase volume buying. Thus you might get the first discount for 100 items, a second for more up to 50, and a third for any more over that. The discount series will appear as 20/10/5, for example, for discounts of 20%, then 10%, and finally 5%.

To apply a series of discounts you treat them as three single discounts on the successive net prices. So

*Net price 1 = List Price * (1- Discount Rate 1)*

*Net price 2 = Net Price 1 * (1- Discount Rate 2)*

*Net price 3 = Net Price 2 * (1- Discount Rate 3)*

You can set up a spreadsheet to calculate the details of the purchase using the above. This gives you the advantage of being prepared to deal with two discounts. You might also set up a spreadsheet to compute the Net Decimal Equivalent and then use its complement to find the final net price. The formulas for the equivalent and then the net price are:

*Net Decimal Equivalent = (1-discount rate1) * (1-discount rate2)*(1-discount rate3)*

*Net price = List Price * (1+Net Decimal Equivalent)*

From a spreadsheet's point of view, both templates are as much work to make, so you might as well show the work and make a worksheet for the first method. However, with the Net Decimal Equivalent you can also find a comparable Single Discount Equivalent, which is the percentage it would take to match the discount series' savings.

Single Discount Equivalent = 1 - Net Decimal Equivalent

*Single Discount = List Price * Single Discount Equivalent*

Exercise 3

Young's Appliance Store offers a trade discount series of 15/10/5 for purchases of 20 or more refrigerators. What is the discount? What is the net price of 20 refrigerators?

1. Go to a new section of the TRADE worksheet.

2. Create the heading **Trade Discount Series**.
3. Skip two rows and enter the following labels down the column: **List Price per Item, Number of Items, Total Cost,** *blank row,* **Discount Rate 1, Discount Rate 2, Discount Rate 3,** *blank row,* **Net Decimal Equivalent,** *blank row,* **Amount Saved, Single Discount Equivalent**.
4. Enter the given data.
5. Above the $600.00 entry and in the column to the right, enter the final label, **Net Price**.
6. Format a line under the Number of Items value and enter a formula for the Total Cost: List Price * Number of items, which in our example is **=F14*F15**.
7. Find the net price for Discount 1 by entering the single step formula: =Total Price * (1-Discount Rate), which in our example is **=F16*(1-F18)**.
8. Use this first Net Price to find the second net price. In our example this is **=G18*(1-F19)**.
9. Copy the second net price formula down one cell to find the final net price.
10. Enter the formula for the Amount Saved which is the difference between the Total Price and the final Net Price. In the example this is **=F16-G20**.
11. Find the Net Decimal Equivalent. In our example this is **=(1-F18)*(1-F19)*(1-F20)**.
12. Next to the Net Decimal Equivalent, enter the alternative formula for Net Price. In our example this is **=F16*F22**.
13. Enter the formula for the Single Discount Equivalent. In the example this is **=1-F24**.
14. Enter the formula for the amount saved using the Single Discount Equivalent. In our example this is **=F16*F25**.

    Note: You could eliminate the Net Decimal Equivalent and Single Discount Equivalent entries without losing any data.
15. Format the Trade Discount Series as shown in Figure 11-2, select it and give it the TRADEDISSERIES range.
16. Save the spreadsheet.

E11 | Trade Discount Series

	E	F	G	H
11	**Trade Discount Series**			
12				
13			**Net Price**	
14	**List Price per Item**	$ 600.00		
15	**Number of Items**	20		
16	**Total Cost**	$ 12,000.00		
17				
18	**Discount Rate 1**	15%	$ 10,200.00	
19	**Discount Rate 2**	10%	$ 9,180.00	
20	**Discount Rate 3**	5%	$ 8,721.00	
21				
22	**Amount Saved**		$ 3,279.00	
23				
24	**Net Decimal Equivalent**	72.7%	$ 8,721.00	
25	**Single Discount Equivalent**	27.3%	$ 3,279.00	
26				

Figure 11-2: Manufacturers and Wholesalers use a Trade Discount Series to sweeten the deal and promote larger volume purchases by retail merchants.

Exercise 4

Compare a deal matching the same number of refrigerators at the same price with trade discounts of 20/5. Which is the better deal?

Multiple Items with Trade Discounts

The above examples make the trade discount problem seem simple and they illustrate the basic math principles you must use when computing discounts. However, a more realistic example would include different discount rates for different items and several different items. Your template must prepare for this common situation. It must also be prepared to handle one, two, or three discounts.

To handle this situation, you can use three IF statements to prepare for one, two, or three discounts. There are not often more than three offered. If you do have such a situation, add a new discount rate column and include a fourth IF statement for computing the Net Decimal Equivalent. The formula we will use is:

Net Decimal Equivalent = IF(Discount1="",then "",otherwise IF(Discount2="",
then 1-Discount1,otherwise IF(Discount3="",(1-Discount1)(1-Discount2),*
otherwise (1-Discount1)(1-Discount2)*(1-Discount3))))*

As a manufacturer you can use this spreadsheet to complete your invoices. As a retail buyer, you can use it to comparison shop between manufacturers and to check the invoice figures.

Exercise 5

The Uni-Source Corp. prepared the invoice in Figure 11-3. Create a spreadsheet to account for multiple items with multiple discounts.

INVOICE

Uni-Source Corp. Paper Supplies
Windsor, MA

BILL TO:	SHIP TO:
May's Stationary Store 1118 Airway Court Santa Rosa, CA 95403	May's Stationary Store 1118 Airway Court Santa Rosa, CA 95403

ORDER DATE: 03/04/96 REFERENCE NO.: SALESPERSON: Terry
SHIP DATE: 03/04/96 SHIP VIA:

ITEM NUMBER	Description DESCRIPTION	OTY ORDER	OTY SHIP	UNIT PRICE	PRICE EXTENSION
20084	PENS	300	300	0.30	
25258	LEGAL PADS	200	200	0.60	
10040	BOX PAPER CLIPS	100	100	0.90	
				SUB-TOTAL	
				TAX 7.5%	
				SHIPPING	
				TOTAL	

Figure 11-3: Use this invoice to complete Exercise 5.

1. In another section of the TRADE worksheet, enter the title **Multiple Item Trade Discount.**

2. Skip a row and enter the column labels: **Item Number, Description, QTY Order, QTY Ship, Unit Price, Discount Rate 1, Discount Rate 2, Discount Rate 3, Net Decimal Equivalent, Price Extension**

3. Fill in the given data from the Uni-Source invoice (Figure 11-3).

4. Enter the Net Decimal Equivalent formula. In our example it is =IF(AB30="","", IF(AC30="",1-AB30,IF(AD30="",(1-AB30)*(1-AC30),(1-AB30)*(1-AC30)*(1-AD30))))

5. Enter the formula for the Price Extension: =(QTY Ship*Unit Price)*(1-Net Decimal Equivalent), which in our example is =(Z30*AA30)*(1-AE30).

6. Copy the formulas down the rows.

7. Two columns before the Price Extension column, enter the row labels: *blank row,* **Sub-Total, Tax, Shipping Charges,** *blank row,* **Total.**

8. Enter a SUM function to total the Price Extension column including the extra row to allow for expansion. In our example it is =SUM(AF30:AF33).

9. Enter 7.5% next to the Tax and enter the formula for computing the tax: =Sub-Total*Tax Rate, which in our example is =AF34*AE35.

10. Enter a SUM function to total the Sub-Total, Tax, and Shipping, including the blank row to allow for other charges.

11. Select the range of cells used and name the range MULTITRADE.

12. Format the range as you see in Figure 11-4 and save the spreadsheet.

W27 | Multiple Item Trade Discount

	W	X	Y	Z	AA	AB	AC	AD	AE	AF
27		Multiple Item Trade Discount								
28										
29	*Item Number*	*Description*	*QTY Order*	*QTY Ship*	*Unit Price*	*Discount Rate 1*	*Discount Rate 2*	*Discount Rate 3*	*Net Decimal Equivalent*	*Price Extension*
30	20084	Pens	300	300	$ 0.30	12%	10%	5%	75%	$ 22.28
31	25258	Legal Pads	200	200	$ 0.60	10%			90%	$ 12.00
32	10040	Paper Clips	100	100	$ 0.90	5%	5%		90%	$ 8.78
33										
34								Sub-Total		$ 43.06
35								Tax	7.5%	$ 3.23
36								Shipping Charges		
37										
38								Total		$ 46.29
39										

Figure 11-4: Multiple items with multiple discounts makes trade discount computations a little more challenging.

Application Exercises

Use the single and series trade discount templates to solve the following problems. Tax remains at 7.5%. When asked, find the discount amount.

1. Montgomery Fashion's ships 1,300 dresses at $4.50 each, 2,200 scarves at $0.18 each, and 1,000 boxes of barrettes at $1.90 each. The trade discount rate for dresses is 12%; for scarves is 5/3; and for barrettes is 5/5/3. What is the total cost of the goods?

2. What are the discount and the net price for 20 boxes of computer paper at $14.67 each and a discount rate of 20%.

3. What is the discount and net price for thirty cases of antifreeze coolant. Each case has six one-gallon units selling for $2.18 per gallon. You get an 18% discount.

4. One manufacturer lists a desk at $700 with a discount series of 20/10/10. A second manufacturer lists the same desk at $650 with a discount series of 10/10/10. Which is the better deal? Why?

5. Dianna Beulke manages an electronics equipment store and has ordered 100 color TVs with remote controls for a special sale. The list price for each TV is $210 with a trade discount series of 7/10/5. What is the discount and net price?

6. Robert Armstrong is purchasing an order of computers for his computer store. He has ordered 36 computers at $1,599 each and a trade discount series of 5/5/10. What are the discount and net price?

7. Cynthia Calhoun is calculating the trade discount on a 6' X 10' X 10' dog kennel that has a list price of $269 with a trade discount series of 10/10/10. What are the discount and net price?

8. Christy Hodge manages a computer software distributorship and offers a desktop publishing software package for $395 with a trade discount series of 5/5/8. What are the discount and net price?

9 One distributor lists ink jet printers with 360 dpi and 6 scaleable fonts that can print envelopes, labels, and transparencies for $189.97 with a trade discount series of 5/5/10. Another distributor lists the same brand and model printer at $210 with a trade discount series for 5/10/10. Which is the better deal if all other factors (shipping, availability, warranty, and the list) are equal?

10. Two distributors offer the same brand and model 486SX2/50 MHz PC with 8 MB RAM, 540 MB hard drive, 3.5" floppy drive, CD-ROM drive, 1 MB video card, and manufacturer's warranty and toll-free help number. One of the distributors lists the computer at $1,899 with a trade discount series of 8/8/5 and the distributor will pay for shipping. The other distributor offers the computer at $2,000 with a trade discount series of 10/5/5 and $50 shipping cost. The shipping cost is added to the net price. Which computer is the better deal?

11. A television set is listed at $400 less 20%. The same set is listed by another manufacturer for $425 less 24%. Which is the better deal?

12. A hutch is listed at $650 with a trade discount of $65. The same hutch is listed by another manufacturer for $595 with a trade discount of 25%. Which is the better deal?

13. One manufacturer lists an aquarium for $58.95 with a trade discount of 10%. Another manufacturer lists the same aquarium for $60 with a trade discount of $9.45. Which is the better deal?

14. One manufacturer lists picture frames at $20 each, discounted 10/10/10. Another manufacturer lists the same picture frames at $19 with a trade discount series of 1/5/10. Which is the better deal?

15. A trade discount series for 10/20 is offered on 20 dart boards that are listed at $14 each. Also, a trade discount series of 20/10 is offered on 10 bowling balls that are listed at $40 each. Find the discount and the total net price for the dart boards and bowling balls.

12. Cash Discounts

Manufacturers and distributors offer trade discounts to get you to buy their products in larger quantities and then offer cash discounts to encourage your early payment of their invoices. Cash discounts come with conditions that tell you when to pay to take advantage of the discount. You can benefit from three possible deals: the Ordinary Dating Discount, the End-of-Month Discount, and the Receipt of Goods (ROG) Discount. Sometimes you are offered sets of discounts. In addition, you must take into consideration the freight charges, which can be Free on Board (FOB) shipping point, FOB destination, or Pay and Add. Finally, your spreadsheet templates must also plan for the discount for partial payments.

Objectives

By working through and completing the exercises below, you will be able to:

- Create a template to compute an Ordinary Dating cash discount.
- Create an End-of-Month cash discount template.
- Create a Receipt of Goods (ROG) cash discount template.
- Include discount sets in your template.
- Include freight charges in your template.
- Include conditions to account for partial payments in your template.

Discount Terms

Ordinary Dating Discount

Invoices provide a breakdown of purchases by item and quantity. They also use a short-hand method to record the payment terms. The most common are the Ordinary Dating Discount terms. The Ordinary Dating Discount includes the discount rate and the number of days in which to take advantage of the discount. For example, the terms 2/10 n/30 reads, "A 2% discount may be deducted from the total price if the invoice is paid on or before 10 days after the date of the invoice. Otherwise, the total price is due in 30 days." So, for an invoice dated April 9 with terms of 2/10, if paid on any day from April 9 to and including April 19, you may deduct 2% of the total price.

You will recognize the cash discount as yet another application of the percentage math problem. In this case you only need to know the terms, multiply the total price by the discount rate, and deduct the product from the total price to find the Net Amount Due. The process can done in two steps or in a single step:

$$Discount = Total\ Price * Discount\ Rate$$

$$Net\ Amount\ Due = Total\ Price - Discount$$

OR

$$Net\ Amount\ Due = Total\ Price - (Total\ Price * Discount\ Rate)$$

You can use your knowledge of adding and subtracting dates to have the spreadsheet work out whether you are paying within the discount period. An IF function can apply or not apply the discount appropriately.

Exercise 1

An invoice dated July 8 shows a net price of $450 with the terms 2/10, n/30. What is the amount due if paid on July 11.

1. Open a new spreadsheet or start a new worksheet and name it CASHDISC.
2. Enter the title, **Cash Discount** and the subtitle **Ordinary Dating Terms**.
3. Skip a row and enter the row labels: **Net Price,** *blank row,* **Discount, Discount Period, Due in,** *blank row,* **Invoice Date, Discount Date, Payment Date, Payment,** *blank row,* **Discount,** *blank row,* **Net Amount Due.**
4. Enter the Net Price and Payment.
5. Enter the Discount Rate, Discount Period, Due in, Invoice Date, and Payment Date.
6. Enter the formula for finding the Discount Date: =Invoice Date + Discount Period, which in our example is =B7+B10.
7. Enter an IF function to determine the Discount: =IF(AND(Payment>0,Payment Date<=Discount Date),then Payment*Discount Rate,otherwise 0), which in our example is =IF(AND(C13>0,B12<=B11),C13*B6,0).
8. Subtract the discount from the Net Price to find the Net Amount Due. In our example this is =C13-C15.
9. Select the range of cells used and give it the DISCOUNTORDINARY name.
10. Format the range as you can see in Figure 12-1 and save the spreadsheet.

A1 | Cash Discount

	A	B	C	D
1	Cash Discount			
2	Ordinary Dating Terms			
3				
4	Net Price		$ 875.00	
5				
6	Discount Rate	2%		
7	Discount Period	10		
8	Due in	30		
9				
10	Invoice Date	8-Jul		
11	Dicount Date	18-Jul		
12	Payment Date	11-Jul		
13	Payment		$ 450.00	
14				
15	Amount Credited		$ 9.00	
16				
17	Net Amount Due		$ 441.00	
18				

Figure 12-1: Once you control for the discount date, a single Ordinary Date cash discount is a simple percentage problem.

End of Month Discounts

Another common term for cash discounts includes the EOM condition. Thus, 2/10 EOM means, "If you pay within 10 days after the end of the month, you can deduct 2% of the net price. The net price is otherwise due at the end of the next month." So an invoice with the terms, 2/10 EOM dated July 8 allows a 2% discount between July 8 up to and including August 10; otherwise you must pay the entire amount by August 31.

Another consideration in the EOM term is the 26th rule. If an invoice is dated before the 26th day of the month the End-of-Month is for the current month. If, dated on or after the 26th, then the End-of-Month refers to the next month, which extends the discount period into the month-after-next. You will control for this with an IF function that uses the DAY function to convert a date to a single-day value, the MONTH function to convert a date to a month value, and the VLOOKUP function to lookup a date value in a table.

Some spreadsheets have a function determining the end of the month. We, however, will first create an End-of-Month table and use the VLOOKUP function to find the appropriate figure to use.

Exercise 2

Newman, Inc., received a bill for cleaning services dated Apr 27 for $5,000 with terms 2/10 EOM. The invoice was paid on Jun 2. What was the due date and was there a discount?

Create the Title and Labels

1. Go to a new section of the CASHDISC worksheet.
2. Create the title **Cash Discount** and the subtitle **End-of-Month**.
3. Skip a row and enter the following labels down the column: **Net Price**, *blank row,* **Discount Rate, Discount Period**, *blank row,* **Due in, Invoice Date**, *blank row,* **Adjusted Invoice Date, EOM**, *blank row,* **Discount Date, Payment Date, Payment**, *blank row,* **Discount**, *blank row,* **Net Amount Due.**
4. Enter the Net Price, Discount Rate, Discount Period, Due in (30), and Invoice Date.

Create an End-of-Month Table

1. Go to another area of the spreadsheet below and to the right of the above.
2. Enter the title **End-of-Month Table**.
3. Enter column labels of **Month, EOM, Days.**
4. Fill in the table as you can see in Table 12-1.

   Table 12-1

End of Month		
Month	**EOM**	**Days**
1	1/31/96	31
2	2/28/96	28
3	3/31/96	31
4	4/30/96	30
5	5/31/96	31
6	6/30/96	30
7	7/31/96	31
8	8/31/96	31
9	9/30/96	30
10	10/31/96	31
11	11/30/96	30
12	12/31/96	31

5. Select the range from the number (Month) 1 to (Days) 31 and name it EOM.

Complete the End-of-Month Discount Worksheet

1. For the Adjusted Invoice Date enter an IF function to take the 26th Rule into account: =IF(DAY(Invoice Date<26), then Invoice Date, otherwise Invoice Date+VLOOKUP(MONTH(Invoice Date),EOM,3)), in our example this is =IF(DAY(H56<26),H56,H56+VLOOKUP(MONTH(H56),EOM,3))).
2. To find the EOM value, use the VLOOKUP function =VLOOKUP(MONTH(Adjusted Invoice Date),End-of-Month ,Column Number), which in our example is =VLOOKUP(MONTH(H58),EOM,2).
3. Find the discount date =EOM + Discount Period, which in our example is =H59+H53.
4. Enter the payment date.
5. Enter the IF function for determining if there will be a discount and how much: =IF(Payment Date<=Discount Date,then Payment*Discount Rate,otherwise 0), which in our example is =IF(H62<=H61,I50*H52,0).
6. Enter the Net Amount Due formula. In our example it is =I50-I65.
7. Give the range of cells used the DISCOUNTEOM name.
8. Format the range as you see in Figure 12-2 and save the spreadsheet.

G47 Cash Discount

	G	H	I	J	K	L
47	Cash Discount					
48	End-of-Month					
49						
50	**Net Price**		$ 5,000.00			
51						
52	**Discount Rate**	2%				
53	**Discount Period**	10				
54						
55	**Due In**	30				
56	**Invoice Date**	27-Apr				
57						
58	**Adjusted Invoice Date**	27-May				
59	**EOM**	31-May				
60						
61	**Dicount Date**	10-Jun				
62	**Payment Date**	2-Jun				
63	**Payment**		$ 5,000.00			
64						
65	**Discount**		$ 100.00			
66						
67	**Net Amount Due**		$ 4,900.00			
68						
69					**End of Month Table**	
70						
71				**Month**	**EOM**	**Days**
72				1	1/31/96	31
73				2	2/28/96	28
74				3	3/31/96	31
75				4	4/30/96	30
76				5	5/31/96	31
77				6	6/30/96	30
78				7	7/31/96	31
79				8	8/31/96	31
80				9	9/30/96	30
81				10	10/31/96	31
82				11	11/30/96	30
83				12	12/31/96	31

Figure 12-2: EOM terms compute the discount period from the last day of the month. You must control for the various end of month dates for each month as well as the 26th rule.

Receipt-of-Goods (ROG) Cash Discount

The Receipt-of-Goods cash discount computes the discount date starting from the receipt of the goods rather than the date the invoice was created. For example, 2/15 ROG means that you may deduct 2% of the Net Price 15 days after you receive the bill. Thus an invoice dated September 6, but for which you did not receive the products or services until September 25, would have a discount period from September 25 until October 10.

Exercise 3

An invoice for machine parts has a net price of $400 and is dated November 9. Sales terms are 2/10 ROG. The machine parts arrive November 13. If the bill is paid on November 21, what is the net amount due?

1. Go to a new section of the CASHDISC worksheet.
2. Create the title **Cash Discount** and subtitle **Receipt-of-Goods.**
3. Skip a row down and enter the following labels: **Net Price,** *blank row,* **Discount Rate, Discount Period,** *blank row,* **Due in, Date of Receipt,** *blank row,* **Discount Date, Payment Date, Discount,** *blank row,* **Net Amount Due.**
4. Enter the given data from the problem statement above.
5. For the Discount Date enter the formula =Date of Receipt + Discount Period, which in our example is =P132+P129.
6. Determine if there will be a discount and how much by entering the Discount formula. In our example it is =IF(P135<=P134,Q126*P128,0).
7. Find the net due amount. In our example this is =Q126-Q136.
8. Name the range of cells used DISCOUNTROG.
9. Format the range as you can see in Figure 12-3 and save the spreadsheet.

0123 | Cash Discount

	O	P	Q	R
123	**Cash Discount**			
124	**Reciept-of-Goods (ROG)**			
125				
126	**Net Price**		$ 400.00	
127				
128	**Discount Rate**	2%		
129	**Discount Period**	10		
130				
131	**Due in**	30		
132	**Date of Receipt**	13-Nov		
133				
134	**Discount Date**	23-Nov		
135	**Payment Date**	21-Nov		
136	**Discount**		$ 8.00	
137				
138	**Net Amount Due**		$ 392.00	
139				

Figure 12-3: The Receipt-of-Goods term computes the discount period from when you receive the goods rather than from the invoice date.

Special Discount Situations

Complications occur in the above formulas for three possible reasons. First you may be able to take different discounts under the terms 2/10 1/15 n/30 and the like. In this case if you cannot pay before the first 10-day period you can still take a discount of 1% up to and including the 15th day. To handle this on your template, of course, all you need to do is replace the second set of terms for the first.

Another consideration is preparing for the payment of freight charges. If the freight charges are marked FOB shipping point, it means the manufacturer or distributor did not pay the shipping charges and you pay the shipping firm upon delivery. In this case the shipping charges will not be on your invoice. If the shipment was FOB destination, then the shipper paid the shipping charges and did not include them in your invoice. However, if the bill of lading is marked Prepay and Add, the shipping charges are included in the Net Price. You cannot take a discount from the freight changes, so your Net Price must be adjusted and the freight charges need to be added to the Outstanding Balance:

Adjusted Net Price = Net Price - Shipping Charges

Outstanding Balance =Net Price - Discount + Shipping Charges

A third possibility complicating the discount spreadsheets is partial payments. If you pay part of the bill within the discount period you can get a partial discount. The amount credited then becomes the payment divided by the complement of the discount rate *(1-Discount Rate)*. To handle this you will have to modify the Discount formula to include AND statements in the IF function statement and label it the Amount Credited. It will look like this:

Amount Credited =IF(AND(Payment Date<Discount Date,Payment<Adjusted Net Price),
then Payment / (1-Discount Rate)),otherwise IF(Payment Date<Discount Date,
*then Net Price*Discount Rate,otherwise 0))*

After you subtract the payment and discount from the Net Price you will have an Outstanding Balance. So:

Outstanding Balance = Net Price - Amount Credited + Freight Charges

You will add the freight charge and partial payment aspects to your spreadsheet and then you will be finished with the template.

Exercise 4

The Semmes Corporation received an $875 invoice dated July 27 for cardboard cartons with terms of 3/10, n/30. The firm could not pay the entire bill within ten days but sent a check for $500 on August 2. Included in the price are freight charges of $95 Pay and Add. What was the outstanding balance?

1. Open the CASHDISC worksheet.
2. Go to the DISCOUNTORDINARY range.
3. Enter the new given data from the problem statement.
4. Change the Discount label to read **Amount Credited**.
5. Insert three rows above the Net Amount Due row.
6. Add the labels **Freight Terms** and **Freight Charges** in the column a row below the Amount Credited.
7. Enter the "Pay and Add" Freight Term.
8. Enter the $95 shipping fee beneath the credit.
9. Insert a row beneath the Net Price.
10. Enter the label **Adjusted Net Price**.

11. Subtract the shipping charges from the Net Price to find the Adjusted Net Price. In our example this is =C4-C19.

12. Edit or rewrite the IF function statement for the discount to include the possibility of partial payments. Our example is =IF(AND(C14>0,B14<=C5),C14/(1-B7), IF(B13<B12,C5*B7,0)).

13. Enter an IF function to find the Outstanding Balance. In our example this is =IF(B18="Pay and Add",C5-C16+C19,C5-C16).

14. Format the table as you can see in Figure 12-4 save your spreadsheet.

A1 | Cash Discount

	A	B	C	D
1	**Cash Discount**			
2	**Ordinary Dating Terms**			
3				
4	**Net Price**		$ 875.00	
5	**Adjusted Net Price**		$ 780.00	
6				
7	**Discount Rate**	3%		
8	**Discount Period**	10		
9	**Due in**	30		
10				
11	**Invoice date**	27-Jul		
12	**Dicount date**	6-Aug		
13	**Payment Date**	2-Aug		
14	**Payment**		$ 500.00	
15				
16	**Amount Credited**		$ 515.46	
17				
18	**Freight Terms**	Pay and Add		
19	**Freight Charges**		$ 95.00	
20				
21	**Net Amount Due**		$ 359.54	
22				

Figure 12-4: Controlling for partial payments and freight charges, your templates should be able to handle most of your buying situations.

Exercise 5

Update the DISCOUNTEOM and DISCOUNTROG ranges to include preparation for Pay and Add freight charges and partial payments.

Application Exercises

Use the appropriate cash discount template to solve for the following.

1. Mr. Williams received an invoice for $200 dated March 6 with terms 1/10, n/30. He paid the bill in full on March 9. What, if any, was his discount and net amount due?

2. Ms. Ryan received a bill for $750 dated September 1 with sales terms of 3/10, 1/15, n/30. What percent discount will she receive if she pays the bill on September 6? What is the discount, if any, and the net amount due?

3. Mrs. Montgomery received a bill for $300 dated April 7. The sales terms on the invoice were 2/10 EOM. If she paid the bill on May 2, how much did Mrs. Montgomery pay?

4. An invoice for $400 dated December 7 has sales terms of 2/10 ROG. The merchandise arrived December 11. If the bill is paid on December 18, what is the net amount due?

5. If the bill in problem 4 is paid on January 2, what is the amount due?

6. The Dean Specialty Company purchased monogrammed items worth $895 and made a partial payment of $600 on day 12. If the sales terms were 3/15, n/30, how much was credited to the account? What was the outstanding balance?

7. The monogrammed items purchased by Dean Specialty Company are shipped by rail from the manufacturer. The bill of lading is marked, "FOB destination." Who pays the freight charges?

8. Mr. Matthews received a bill dated March 1 with sales terms of 3/10, n/30. What percent discount will he receive if he pays the bill on March 5? What is the discount and net amount due?

9. Ms. Wagner received a bill dated September 3 with sales terms of 2/10, n/30. Did she receive a discount if she paid the bill on September 15?

10. Campbell Sales purchased merchandise worth $745 and made a partial payment of $300 on day 13. If the sales terms were 2/15, n/30, how much was credited to the account? What was the outstanding balance?

11. Elmer Wilson purchased roll vinyl floor covering for $1,150 with sales terms of 3/15, n/30. He paid $800 on the bill within 15 days. How much was credited to his account? What was his outstanding balance after the payment? How would this be different if shipping charges of $150 were Pay and Add? (For part 2 add the shipping charges to the Net Price.)

12. An invoice for $1,200 is dated June 3, and terms of 3/10, n/30 are offered. A payment of $800 is made on June 12, and the remainder is paid on July 2. Find the amount remitted on July 2 and the total amount paid.

13. Find the amount credited and the outstanding balance on an invoice dated August 19 if a partial payment of $500 is paid on August 25 and has terms of 3/10, 1/15, n/30. The amount of the invoice is $826, which includes $87 in shipping charges Pay and Add.

14. How much would have to be paid on an invoice for $28 with terms of 3/10 EOM if the bill dated June 8 is paid July 2? How much is due if it includes a $6 shipping and handling charge?

15. How much would have to be paid on an invoice for $328 with terms of 2/10 ROG if the merchandise invoice is dated January 3 and the merchandise arrives FOB shipping point on January 8? $150 is paid on January 11. What, if any, discount is allowed? What, if any, shipping charge is included? What is the final amount paid?

13. Inventory and the Cost of Goods Sold

The cost of goods sold, which is a concept you used in Chapter 3 on Financial Statements, is related to the inventory method a business uses. It is the cost of goods available for sale (Cost of Beginning Inventory plus Purchases) minus the cost of ending inventory. It is a record of the cost of merchandise sold for a given time period.

In this chapter you will learn about three different inventory methods and use two of them. Keep in mind that inventory methods differ from business to business. This chapter does not attempt to show you how to use the spreadsheet to conduct an inventory or keep data on an inventory. That is possible but more efficiently done using a data base program from which you can extract the data needed to compute the cost of goods sold.

Objectives

By working through and completing the exercises below, you will be able to:

- Briefly explain the Cost of Goods Sold Specific Identification inventory method.
- Briefly explain the FIFO and LIFO inventory methods.
- Prepare a Cost of Goods Sold Weighted-Average inventory spreadsheet template.
- Prepare a Cost of Goods Sold Retail inventory spreadsheet template.
- Include the Turnover Rate at Cost in your templates.
- Include the Turnover Rate at Retail in your templates.

Specific Identification, FIFO, and LIFO Inventory Methods

Three inventory methods not conducive to spreadsheet use are the Specific Identification, FIFO (First-in-First-Out) and LIFO (Last-In-First-Out) methods. Though it is not beyond the realm of possibility to use a spreadsheet with these inventory methods, it is an unlikely scenario. The Specific Identification method needs the support of a database program and the other two usually come as a specialized accounting software module.

The specific identification inventory method attempts to keep track of when merchandise was purchased and at what price. For example, for an automobile dealership with sales on different cars coming in at different prices, the specific identification method keeps track of each individual car and its price. This is the most accurate approach to inventory management but also the most difficult to manage. Use of the specific identification inventory method is usually limited to big ticket items like automobiles. With very large inventories of smaller items it can be overly cumbersome.

Many business people want to know how much it will cost to replace existing stock. To get this estimate, instead of using the exact cost of each item in stock to calculate the Cost of Goods Sold, they use the First-in First-out (FIFO) inventory method. This method assumes that the first items brought into the store are the first ones sold and therefore the latest price serves as the price used in computing the cost of goods available and the ending inventory.

For a clearer understanding of the real value of the business' stock, the Last-in, First-out (LIFO) method assumes what stock is left in inventory is the oldest items purchased. Usually the newer items cost more and the LIFO method shows an over-all lower income level than the FIFO method. This allows the business to understand it true worth and report lower income for tax purposes. The cost of the goods available for sale and the replacement costs for new goods could be significant, but the LIFO Cost of Goods Sold should closely predict the cost of replacement.

Weighted-Average Inventory Method

The Weighted-Average method for computing the Cost of Goods Sold starts by finding the average cost for each kind of inventory item. For example, a stationary store might find the average cost per greeting cards, the average cost per pen sets, and the average cost per knick-knack. It then multiplies these figures times the inventory count and adds the results to find the total cost of ending inventory. The formula for this single value is:

Average Unit Cost = Cost of Goods Available for Sale / Units of Goods Available for Sale

Next the Weighted-Average method finds the cost of the ending inventory:

*Cost of Ending Inventory = Units Ending Inventory * Average Unit Cost*

Finally, the Cost of Goods Sold formula remains the same.

Cost of Goods Sold = Cost of Beginning Inventory +Cost of Purchases - Cost of Ending Inventory

Exercise 1

Prepare a Weighted-Average Inventory to find the Cost of Goods Sold for the information in Table 13-1.

Table 13-1

Purchases Date of Purchase	Units Purchased	Cost per Unit	Total Cost
Beginning Inventory	29	$ 8.00	
15-Jan	11	$ 7.00	
4-Feb	9	$ 10.00	
3-Mar	14	$ 8.00	
Goods Available for Sale			
Units Sold			
Ending Inventory	22		
Cost of Goods Sold			

1. Create the INVENTRY worksheet.
2. Enter the title **Cost of Goods Sold** and subtitle **Weighted-Average Inventory Method**.
3. Skip two rows and enter the labels you can see in Table 13-1.
4. From Table 13-1, provide the Units Purchased and the Units Ending Inventory
5. Find the goods available for sale. In our example this is =SUM(BD234:BD238).
6. Enter the formula for the Beginning Inventory's Total Cost. In our example it is =BD234*BE234.
7. Copy the Total Cost formula down to the 3-Mar row.
8. Enter a formula to find the total cost of Goods Available for Sale. In our example it is =SUM(BF234:BF238).
9. Between the Units available and the Cost of the available, enter the formula for the Average Unit Cost. In our example it is =BF239/BD239.
10. Enter a formula to find the number of Units Sold: Goods Available for Sale - Ending Inventory, which in our example is =BD239-BD241.
11. Enter the formula for the cost of Ending Inventory. In our example it is =BF239*BD241.
12. Enter the formula for Cost of Goods Sold. In our example this is =BG239-BG241.

13. Select the range and give it the COGSWEIGHTED name.
14. Format the table as you can see in Figure 13-1 and save the spreadsheet.

BC228 | Cost of Goods Sold

	BC	BD	BE	BF	BG
228		**Cost of Goods Sold**			
229	Weighted-Average Inventory Method				
230					
231	**Purchases**				
232	**Date of Purchase**	**Units Purchased**	**Cost per Unit**	**Total Cost**	
233					
234	Beginning Inventory	29	$ 8.00	$ 232.00	
235	15-Jan	11	$ 7.00	$ 77.00	
236	4-Feb	9	$ 10.00	$ 90.00	
237	3-Mar	14	$ 8.00	$ 112.00	
238					
239	**Goods Available for Sale**	63	$ 8.11	$ 511.00	
240	**Units Sold**	41			
241	**Ending Inventory**	22		$ 178.44	
242					
243	**Cost of Goods Sold**			$ 332.56	
244					

Figure 13-1: The Weighted-Average method of inventory makes computations easier but is only accurate during times of relative economic stability.

Retail Inventory Method

When you just cannot do a physical inventory on a monthly or other periodic basis, you can use an estimate of the cost of inventory. The Retail method uses a ratio comparing the cost of goods available for sale to the retail price of those goods. In doing so you must refer to the cost of ending inventory as the "ending inventory at cost" and the retail price of ending inventory as the "ending inventory at retail."

This method requires formulas for the Cost Ratio, Ending Inventory at Retail, Ending Inventory at Cost, and a new formula for the Cost of Goods Sold:

Cost Ratio = Cost of Goods Available for Sale / Retail Value of Goods Available for Sale

Ending Inventory at Retail = Retail Price of Goods Available for Sale - Sales

*Ending Inventory at Cost = Ending Inventory at Retail * Cost Ratio*

*Cost of Goods Sold = Sales * Cost Ratio*

Exercise 2

Find the Cost of Goods Sold for the data in Table 13-1 using the Retail method. Also use the information in Table 13-2.

Table 13-2

Retail Prices	
Date of Purchase	**Retail Price per Unit**
Beginning inventory	$ 11.41
15-Jan	$ 16.36
4-Feb	$ 14.33
3-Mar	$ 14.33
Sales	$ 487.00

1. Open the INVENTRY worksheet.
2. Go to a new area of the spreadsheet and enter the title **Cost of Goods Sold** and the subtitle **Retail Inventory Method**.
3. In the first column and one row below the subtitle, enter the following row labels: **Date of Purchase, Beginning Inventory, 15-Jan, 4-Feb, 3-Mar,** *blank row,* **Goods Available for Sale,** *blank row,* **Cost of Goods Available for Sale,** *blank row,* **Sales,** *blank row,* **Cost Ratio,** *blank row,* **Ending Inventory at Retail,** *blank row,* **Ending Inventory at Cost,** *blank row,* **Cost of Goods Sold**
4. Add column labels across the Date of Purchase row, enter **Units Purchased, Retail Price per Unit, Retail Price**
5. Enter the given values from the COGSWEIGHTED range (including the Cost of Goods Available for Sale) and Table 13-2 above.
6. Enter a formula to find the first Retail Price: Units Purchased * Retail Price per Unit, which in our example is =Y109*Z109.
7. Copy the Retail Price formula down the next three rows.
8. For the Goods Available for Sale formula, sum the retail prices. In our example this is =SUM(Z109:Z113).
9. In the Retail Price per Unit column, enter the formula for and find the cost ratio. For the example its formula is =Z116/Z114.
10. Find the Ending Inventory at Retail amount. In our example this is =Z114-Z118.
11. Find the Ending Inventory at Cost amount. In our example this is =Z122*Y120.
12. Enter the formula for Cost of Goods Sold. In the example this is =Z118*Y120.
13. Name the used area the COGSRETAIL range.
14. Format the spreadsheet as you can see in Figure 13-2 and save the spreadsheet.

W105 | Cost of Goods Sold

	W	X	Y	Z
105	**Cost of Goods Sold**			
106	**Retail Inventory Method**			
107				
108	**Date of Purchase**	**Units Purchased**	**Retail Price per Unit**	**Retail Price**
109	Beginning Inventory	29	$ 11.41	$ 331.00
110	15-Jan	11	$ 16.36	$ 180.00
111	4-Feb	9	$ 14.33	$ 129.00
112	3-Mar	14	$ 11.43	$ 160.00
113				
114	**Goods Available for Sale**			$ 800.00
115				
116	**Cost of Goods Available for Sale**			$ 511.00
117				
118	**Sales**			$ 487.00
119				
120	**Cost Ratio**		0.64	
121				
122	**Ending Inventory at Retail**			$ 313.00
123				
124	**Ending Inventory at Cost**			$ 199.93
125				
126	**Cost of Goods Sold**			$ 311.07
127				

Figure 13-2: The Retail Method is one of the easiest ways to compute a Cost of Goods Sold figure, but it is only an estimate.

Turnover Rate

Merchandising businesses and any business that uses consumable supplies must watch its inventory turnover rate. With it they can plan to replace stock or supplies well in advance of their demand.

You will learn to compute two useful turnover rates. One is based on the cost of inventory while the second is based on the retail value of the inventory. For both you must use another figure representing the average inventory at cost and the average inventory at retail. The needed spreadsheet formulas are:

Average Inventory at Cost = (Beginning Inventory at Cost + Ending Inventory at Cost) / 2

Turnover Rate at Cost = Cost of Goods Sold / Average Inventory at Cost

Average Inventory at Retail = (Beginning Inventory at Retail + Ending Inventory at Retail) / 2

Turnover Rate at Retail = Sales / Average Inventory at Retail

Exercise 3

Building onto what you have covered so far, continue to work with the Table 13-1 and Table 13-2 data. For this example the Beginning Inventory at Retail is $267.

1. Open the INVENTRY worksheet.
2. Beneath the COGSRETAIL range and to its right, add the title and subtitle: **Turnover Rates for Period, Jan 15 to Mar 3**
3. Skip a row and add labels for the turnover rates: **Beginning Inventory at Cost, Ending Inventory at Cost, Average Inventory at Cost, Turnover Rate at Cost,** *blank row,* **Beginning Inventory at Retail, Ending Inventory at Cost, Average Inventory at Retail, Turnover Rate at Retail**
4. Enter the given data. (Beginning Inventory at Cost can be found in the COGSWEIGHTED range. Ending Inventory at Cost and the Ending Inventory at Retail Re estimated in the in the COGSRETAIL range.)
5. Enter the formula for the Average Inventory at Cost. In the example it is =(AB145+AB146)/2.
6. Enter the formula for the Turnover Rate at Cost two cells over from its label. In the example this is =Z126/AB147.
7. Enter the formula for the Average Inventory at Retail. In the example it is =(AB150+AB151)/2.
8. Enter the formula for the Turnover rate at Retail two cells over from its label. In the example this is =Z118/AC152.
9. Name the used cells the TURNOVERRATES range.
10. Format the range as you can see in Figure 13-3 and save the spreadsheet.

AA142 | Turnover Rates for Period

	AA	AB	AC	AD
142	**Turnover Rates for Period**			
143	Jan 15 to Mar 3			
144				
145	**Beginning Inventory at Cost**	$ 232.00		
146	**Ending Inventory at Cost**	$ 178.44		
147	**Average Inventory at Cost**	$ 205.22		
148	**Turnover Rate at Cost**		1.52	
149				
150	**Beginning Inventory at Retail**	$ 331.00		
151	**Ending Inventory at Retail**	$ 313.00		
152	**Average Inventory at Retail**	$ 322.00		
153	**Turnover Rate at Retail**		1.51	
154				
155				

Figure 13-3: The turnover rates for our example shows that during the period studied, the inventory was sold and replaced twice.

Application Exercises

For the information in each of the following tables, find the Cost of Goods Sold for the given periods. Use the Weighted-Average method and then the Retail Inventory method.

1. Sales equal $78,982.

Purchases				**Retail**	
Date of Purchase	**Units Purchased**	**Cost per Unit**	**Total Cost**	**Retail Price per Unit**	**Total Retail Price**
Beginning Inventory	42	$ 850.00		$ 975.00	
5-Feb	21	$1,760.00		$ 2,115.00	
19-Feb	17	$ 965.00		$ 1,206.00	
3-Mar	28	$ 480.00		$ 600.00	
Goods Available for Sale					
Units Sold					
Ending Inventory	34				
Cost of Goods Sold					

2. Sales equal $1,691.

Purchases				Retail	
Date of Purchase	Units Purchased	Cost per Unit	Total Cost	Retail Price per Unit	Total Retail Price
Beginning Inventory	96	$12.00		$18.00	
12-Apr	23	$ 9.00		$13.00	
8-May	15	$11.00		$17.00	
2-Jun	37	$15.00		$21.00	
Goods Available for Sale					
Units Sold					
Ending Inventory	114				
Cost of Goods Sold					

3. Find the turnover rates for application exercise 1.
4. Find the turnover rates for application exercise 2.

14. Overhead

From buying products and keeping track of their turnover naturally comes the subject of overhead and markup. Markup, which you will work on in the next chapter, prices merchandise and services based on their cost or sales and must recover for what the business pays in expenses. Expenses are also called overhead. In this chapter you will focus on two ways a business can distribute its overhead expenses.

Objectives

By working through and completing the exercises below, you will be able to:

- Prepare a distribution of overhead template based on sales.
- Prepare a distribution of overhead template based on floor space.
- Include the Overhead rate in your templates based on sales.

Overhead Based on Sales

Overhead is the cost for maintaining the building, obtaining and shipping supplies, depreciation, maintenance, utilities, and more. A large percentage of business income goes to maintaining its overhead costs. You will see how this affects prices but for now the problem lies in how to allocate those costs.

The first method for which you will build a template is overhead based on sales. In this spreadsheet, you will distribute the cost of overhead among the different parts of a business, such as a retail store, based on that area's productivity. This information lets you analyze the different areas of your business and allocate your resources in a reasonable manner.

The distribution is based on the department sales rate, which is the department sales divided by the total sales. Multiplying this percentage by the total overhead cost lets you see what portion of overhead is covered by which departments.

Exercise 1

Just for Fun's overhead totaled $8,000 during one month. It has five departments. Base the distribution of overhead on sales if the store had the following monthly sales by department: cameras, $5,000; jewelry, $8,200; sporting goods, $6,700; silver, $9,200; and toys, $12,000.

1. Start a new worksheet and name it OVERHEAD.
2. Enter the title **Overhead Based on Sales**.
3. Skip a row and create the row label **Total Overhead Expenses** and in the cell three rows to the right, enter the total cost of overhead.
4. Skip a row and list the column headings: **Department, Sales, % of Sales,** and **Total Overhead Expense**.
5. Under the Department column and Sales column list the department names and sales figures from the problem statement above. Skip a row and add the **Total Sales** row label.
6. Skip row below the last sales figure and total the sales using the SUM function and be sure to include the extra row at the bottom of the range so your formula will automatically include any rows you may insert. In our example the formula is =SUM(B6:B11).
7. For the first sales rate enter a formula to divide the camera sales by the total sales. In our example it is =B6/$B$12.

8. Under the column heading Total Overhead Expense, multiply the camera sales times the % of Sales. In our example it is **=C6*$B$3**.
9. Copy the sales rate and overhead formulas down to the Toys row.
10. Copy the Total Sales formula to the same row but beneath the Total Overhead Expenses column. In our example the cell is D12.
11. Save the range as OHSALES.
12. Check your work against Figure 14-1 below and save the spreadsheet.

A1 | Overhead Based on Sales

	A	B	C	D	E
1	**Overhead Based on Sales**				
2					
3	Total Overhead Expense			$ 8,000.00	
4					
5	**Department**	**Sales**	**% of Sales**	**Total Overhead Expense**	
6	Cameras	$ 5,000.00	12.17%	$ 973.24	
7	Jewelry	$ 8,200.00	19.95%	$ 1,596.11	
8	Sporting goods	$ 6,700.00	16.30%	$ 1,304.14	
9	Silver	$ 9,200.00	22.38%	$ 1,790.75	
10	Toys	$12,000.00	29.20%	$ 2,335.77	
11					
12	Total Sales	$41,100.00		$ 8,000.00	
13					

Figure 14-1: Overhead by Sales show how much of the overhead is associated with each area of sales.

Overhead Based on a Constant

Another way to look at your overhead costs is to distribute them across some constant characteristic of your business. You might do so for the amount of time each employee and yourself spend producing the income or miles driven for a delivery service. Here we will use the example of a store and distribute the overhead cost across the floor space in the building. This is reasonable for a store since it can indicate how each area of the store contributes to the upkeep of the business.

For this example, you will create the Department floor-space rate which is the floor space in a department divided by the total floor space. Multiplying this percentage by the total cost of overhead gives you the distribution figures.

Exercise 2

The Super Store assigns overhead to its various departments according to the floor space used by each department. The store's total overhead is $25,000. Find the overhead for each department if each department uses the following square feet: junior department, 3,000; women's wear, 4,000; men's wear, 3,500; children's wear, 3,000; china and silver, 2,500; housewares, 2,500; linens, 2,000; toys, 1,500; carpets, 3,500; and cosmetics, 500.

1. Open the OVERHEAD worksheet.
2. Copy the OHSALES range to a new area of the worksheet.
3. Edit the title to read **Overhead by Floor Space**.

4. Enter the Overhead amount from the problem statement above.
5. Edit the Sales and Sales Rate column headings to read **Floor Space in Square Feet** and **% of Floor Space** respectively.
6. Overwrite the Department headings inserting rows as needed.
7. Overwrite the Sales figures with the Floor Space figures. The column should sum as you proceed.
8. Edit the first Floor-Space Rate formula to reference the new floor-space Total cell. Make sure you use an absolute reference. In our example it is **=M43/$M$54**.
9. Edit the first Total Overhead Expense formula to reference the new Overhead amount making sure to use an absolute reference. In our example it is **=N43*$O$40**.
10. Copy the new formulas down the columns. The Total Overhead Expense should sum in the totals row.
11. Select the range of cells used and name it OHFLOOR.
12. Format the table as you can see in Figure 14-2.
13. Save the spreadsheet.

L38 | Overhead Based on Floor Space

	L	M	N	O	P
38	**Overhead Based on Floor Space**				
39					
40	Total Overhead Expense			$ 25,000.00	
41					
42	**Department**	**Floor Space in Square Feet**	**% of Floor Space**	**Overhead**	
43	Junior department	3,000	11.54%	$ 2,884.62	
44	Women's wear	4,000	15.38%	$ 3,846.15	
45	Men's wear	3,500	13.46%	$ 3,365.38	
46	Children's wear	3,000	11.54%	$ 2,884.62	
47	China and silver	2,500	9.62%	$ 2,403.85	
48	Housewares	2,500	9.62%	$ 2,403.85	
49	Linens	2,000	7.69%	$ 1,923.08	
50	Toys	1,500	5.77%	$ 1,442.31	
51	Carpets	3,500	13.46%	$ 3,365.38	
52	Cosmetics	500	1.92%	$ 480.77	
53					
54	Total Floor Space	26,000		$ 25,000.00	
55					
56					

Figure 14-2: Basing Overhead costs on a constant factor reveals the productivity of that constant.

Finding the Overhead Rate

Of special interest to markup and markdown considerations is the cost of overhead. Every business needs to have some way of measuring whether its income is meeting or surpassing its needs. One measure to help them is the Overhead Rate, which is the Total Cost of Overhead divided by total sales or income. Of course, this number has different meaning for every business. It basically points to a ratio between the two figures. If it were 1.00 or 100%, then sales would only be covering Total Overhead Expenses with every dollar of sales going to keep the business open. In a sense this is breaking even; however, overhead does not include the cost of goods. A figure greater than 1.00 shows an operating loss although some businesses occasionally endure periods with this loss since it is offset by later gains.

Exercise 3

Find the Overhead rate for the Just For Fun store in Exercise 1 above.

1. Open the OVERHEAD worksheet and go to the OHSALES range.
2. Beneath the Total Sales label, enter the **Overhead Rate** label.
3. In the cell opposite this label enter the formula for Overhead Rate: =Overhead/Sales, which in our example is =D3/B12. The answer should be 19.46% showing close to 20¢ in overhead being paid by every $1 of sales or 80¢ left for other purposes.
4. Save the spreadsheet.

Application Exercises

Use your overhead templates to analyze the following business data. Write a brief report on each.

1. Overhead for one month at the Allimore Department Store totaled $6,000. Find the overhead for each department, based on sales, if the store had the following monthly sales by department: Toys, $4,000; appliances, $6,600; children's clothing, $6,800; books, $4,600; and furniture, $8.400.
2. Dale Crosby's Gift Shop had overhead expenses totaling $2,732 during the month of August. The business recorded departmental sales for the month of August as follows: china, $3,923; silver, $8,923; crystal, $2,932; linens, $1,923; new gifts, $6,291; antiques, $8,923. Use this information to find the overhead for each department based on sales.
3. Carlisle's Stock Trailer Sales has overhead expenses that totaled $4,932 during one month. The business had the following departmental sales for the month: cattle trailers, $8,523; utility trailers, $6,201; boat trailers, $2,932; parts, $1,392. Find the overhead for each, based on sales.
4. Savemore Discount Clothing Store assigns overhead to its various departments according to the floor space used by each department. The store's total monthly overhead is $15,800. Find the overhead for each department using the following square feet for each: women's clothing, 2,000; men's clothing, 1,200; children's clothing, 2,500.
5. Hughe's Trailer Manufacturer assigns overhead to its departments according to the floor space used by each department. The company's total monthly overhead for the month of April is $7,832. Find the overhead for each department using the following square feet for each department: welding bay, 2.100; paint shop, 1,950; axles and steel storage, 780; flooring lumber, 380; office space, 500.
6. Office Supply World assigns overhead to a department based on the square feet of office space it occupies. The overhead for a month totaled $9,000 and each department occupies the following number of square feet: furniture, 2,000; computer supplies, 1,600; consumable office supplies, 2,500; leather goods, 1,200; and administrative services, 800. Find each department's overhead.
7. AMX Department Store's overhead totaled $12,000 during one month. The sales by department for the month were as follows: cameras, $12,000; toys, $14,000; hardware, $12,500; garden supplies, $8,400; sporting goods, $9,500; and clothing, $28,600. Report on the monthly overhead for all departments.
8. The office photocopy machine is on the blink again. You are responsible for replacing the photocopier with a more powerful model and equitably charging each department its share of the cost of the new copier. The new copier costs $7,580 and is expected to produce 500,000 copies in its lifetime. You decide that each department's share should be based on the number of copies the

department makes. These figures for the last year were: purchasing, 8,711; personnel, 30,872; payroll, 32,521; secretarial pool, 52,896.

9. Department A uses 5,000 square feet of floor space. Department B uses 2,500, Department C uses 4,300, and Department D uses 2,700. The total overhead is $8,200. Find each department's overhead based on floor space.

10. Tyson's Fixit Store has a monthly overhead of $9,200. Find each department's monthly overhead based on floor space using the following square feet for each department: hardware, 800; plumbing, 600; tools, 400; supplies, 600.

11. A corner grocery store has a monthly overhead of $1,500. Find each department's monthly overhead based on sales if department sales are as follows: meats, $1,200; groceries, $2,400; dairy, $600; and housewares, $800.

12. At Best Buy Hardware, the nuts and bolts department had $1,500 in sales for the month, the electrical department had $4,000, and the paint department had $2,300. The total overhead was $3,800. Find each department's overhead based on sales.

13. Department 1 had $5,200 in sales for the month, Department 2 had $4,700, Department 3 had $6,520, Department 4 had $4,870. The total overhead was $10,000. Find each department's overhead based on sales.

14 The operating expense of Freeze Bros., a department store, are allocated according to the total floor space occupied by each department. Find the overhead for each if the May overhead totals $80,000. Each department occupies the following floor space: Women's wear, 6,000; Men's wear, 4,500; children's wear, 3,000; Cosmetics, 1,500; Home furnishings, 5,000. Find the distribution of overhead based on floor space.

15. Benji's Pizza and Takeout distributes its delivery and vehicle maintenance expenses by the number of miles driven per vehicle. The delivery cars had the following mileage for August: Car 1, 3,456.6; Car 2, 2,678.3; Car 3, 1,957.1; Car 4,895.9. How much of the $8,689.97 expenses go to each car?

15. Markup and Markdown

A savvy business person knows what it costs to stay in business, how much profit is reasonable, and what price can be charged for the merchandise or service the business provides. This ability is based on experience, a careful eye on the bottom-line, and some measure of intuition or "business smarts."

In determining markup and markdown, the business person decides at what price to initially offer products for sale and how much to reduce the price as needed as goods go out of style or do not sell in time for new stock. Some businesses still have a tradition of bargaining on the price of goods and as such these business people must stay constantly aware of what they may be sacrificing.

In this chapter you will create templates to assist you in this important decision making process.

Objectives

By working through and completing the exercises below, you will be able to:

- Prepare a Markup Based on Cost template.
- Prepare a Markup Based on Selling Price template.
- Modify the markup template to study and keep track of markdowns.

Markup Based on Cost

With experience, many businesses settle on a markup rate that works for them. Prices then increase as their costs increase. The markup equals the markup rate times the net price of an item. The selling price is then computed by adding the markup to the net price. As you have seen, using the one plus a rate times a base amount will provide a one-step method for computing the target amount. In this case:

*Selling Price = Net Cost * (1 + Markup Rate)*

Exercise 1

Shelby is interested in a new car that cost Marty's Auto Sales $20,000 to put on the showroom floor. Marty's uses a markup rate of 22%. What is the asking price of the car in which Shelby is interested?

1. Open a new worksheet and name it MARKUP.
2. Enter the title **Markup Based on Cost**.
3. Skip a row and enter the row labels: **Item, Markup Rate,** *blank row,* **Cost, Markup, Selling Price (Cost + Markup), Selling Price (One Step).**
4. Enter the known data from the problem statement above.
5. Enter the formula for the Markup: Markup Rate * Cost, which in our example is =B5*B7.
6. Enter the formula for the Selling Price: Cost + Markup, which in our example is =B7+B8.
7. Enter the one-step formula for the Selling Price: Cost * (1+Markup Rate), which in our example is =B7*(1+B5).
8. Select the range and give it the MARKUPCOST name.
9. Format the table as you can see in Figure 15-1 and save the spreadsheet.

A1	Markup Based on Cost	
	A	**B**	**C**

	A	B	C
1	**Markup Based on Cost**		
2			
3			
4	**Item**	Auto	
5	**Markup Rate**	22.0%	
6			
7	**Cost**	$ 20,000.00	
8	**Markup**	$ 4,400.00	
9	**Selling Price (Cost + Markup)**	$ 24,400.00	
10	**Selling Price (One Step)**	$ 24,400.00	
11			
12			

Figure 15-1: Markup is a simple percentage problem and is quickly computed once you know the Markup rate and cost of a product or service.

Markup Based on Selling Price

Some businesses use the markup based on sales because they base everything else on sales. They maintain records that show the average percent of markup based on sales and apply it generally. In such a case the selling price and markup are found by the following formulas:

Selling Price = Cost/(1-Markup Rate)

Markup = Selling Price - Cost

Markup Rate = Markup / Selling Price

Exercise 2

A department store prices its merchandise so as to obtain a gross profit of 24% of the selling price. What selling price will be shown on a redwood picnic table. The invoice for 100 of these tables showed a total cost of $3,856.97.

1. Copy the MARKUPCOST range to a new area of the MARKUP worksheet.
2. Edit the title to read, **Markup Based on Selling Price**.
3. Replace the Markup label with the Selling Price label and vice versus (see Figure 15-2).
4. Insert two rows above the Cost row and two below it.
5. Enter **Number of Items** above the Cost label.
6. Enter the given data from the problem statement.
7. Enter **Cost per Item** below the Cost label and enter its formula in the adjacent cell: =Cost / Number of Items, which in our example is **=J55/J54**.
8. Enter the formula to find the Selling Price. In our example this is **=J56/(1-J52)**.
9. Subtract the Selling Price from Cost to find the Markup: **=J58-J56**.
10. Select the range and give it the MARKUPSALES name.
11. Format the table as you can see in Figure 15-2.
12. Save the spreadsheet.

I48	Markup Based on Selling Price	
	I	**J**	**K**

	I	J	K
48	**Markup Based on Selling Price**		
49			
50	**Item**	Picnic Table	
51			
52	**Markup Rate**	24%	
53			
54	**Number of Items**	100	
55	**Cost**	$ 3,856.97	
56	**Cost per Item**	$ 38.57	
57			
58	**Selling Price**	$ 50.75	
59	**Markup**	$ 12.18	
60			
61			

Figure 15-2: Markup based on sales.

Markdown on Selling Price

A business person must prepare for the fact that at least some inventory will not be sold at the profit margin desired. When you have some knowledge about this rate of loss such as the spoilage rate for a hundred pounds of bananas, you can remove the amount of expected loss from the Number of Items in the spreadsheet. If you know about how many clothing items will not sell as styles change rapidly, you can plan to sell them at a reduced price. The reduction is called markdown.

Most businesses will have a markup rate higher on some items and lower on others. Often some items will be carried at a loss or kept at a break even point while other items make greater profits. The business owner's savvy knowledge about the business' customer and the market determines the desired profit and markup rates.

In our exercise you will find the amount of the selling price that covers overhead using the following formula:

*Overhead Covered by Selling Price = Selling Price * Overhead Rate*

From this you can determine how much of the markup is net profit or loss and the percent of this that the selling price represents (the Percent of Net Profit or Profit Rate):

Net Profit or Loss = Markup - Overhead Covered by Selling Price

Percent of Net Profit = Net Profit or Loss / Selling Price

Given the overhead rate and the desired profit rate, you can create a spreadsheet that tells you when the markdown represents "breaking even." The next level down is one that does not cover the overhead and gives an "Operating Loss." Next is a sale price that does not cover the cost of the goods, an "Absolute Loss."

=IF(Selling Price - Cost<0,then "Absolute Loss",otherwise IF(Net Profit or Loss<0,then "Operating Loss",otherwise IF(Net Profit=0,then "Break Even",otherwise "OK"))),

Exercise 3

Find a comfortable markdown rate for the Picnic tables in Exercise 2. The overhead rate is 9.8% and the department store is comfortable with a 6% rate of profit.

1. Open the MARKUP spreadsheet and go to the MARKUPSALES range.
2. Insert a row above the Markup Rate label.
3. To the right and above the Markup Rate, enter the heading **Markdown Rate** and next to it Sale Price.
4. In the Markdown Rate column and in the Selling Price row, enter 15%.
5. To the right of this, enter a formula to find what 15% of the selling price equals: =Selling Price - (Selling Price * Markdown Rate) or =Selling Price * (1- Markdown Rate). In our example it is =J58-(K58*J58).
6. Below this in the Markup row, subtract the Cost per Item from the Sale Price. In our example this is =L58-J56.
7. Beneath the Markup row, enter new row labels: *blank row,* **Overhead Rate, Overhead,** *blank row,* **Net Profit or Loss, Percent of Net Profit,** *blank row,* **Desired Rate of Profit, Results.**
8. Enter the Overhead Rate from the problem statement as well as the Desired Rate of Profit.
9. For the Net Profit or Loss, enter the difference between the new markup and the Overhead with the overhead reference absolute. In our example this is =J59-$J$62.
10. Enter the formula for Percent of Net Profit. In our example this is =J64/J58.
11. Copy the Net Profit or Loss and Percent of Net profit formulas and the Desired Rate of Profit over to the Sale Price column.
12. Enter an IF formula in the Results row under the Sale Price column. In our example this is =IF(L58-J56<0,"Absolute Loss",IF(L64<0,"Operating Loss",IF(L64=0,"Break Even","OK"))). The results should read "Operating Loss."
13. Copy the Results formula for the Sale Price column to the original column's Results cell. It should read "OK."
14. Change the Markdown Rate to find a variety of possible scenarios. What percentage shows the first Absolute Loss. When would this be OK?
15. Compare your work to Figure 15-3 and save the spreadsheet.

I48	Markup Based on Selling Price			
	I	**J**	**K**	**L**
48	**Markup Based on Selling Price**			
49				
50	**Item**	Picnic Table		
51			**Markdown Rate**	**Sale Price**
52	**Markup Rate**	24%		
53				
54	**Number of Items**	100		
55	**Cost**	$ 3,856.97		
56	**Cost per Item**	$ 38.57		
57				
58	**Selling Price**	$ 50.75	15.00%	$ 43.14
59	**Markup**	$ 12.18		$ 4.57
60				
61	**Overhead Rate**	9.8%		
62	**Overhead Covered by Selling Price**	$ 4.97		
63				
64	**Net Profit or Loss**	$ 7.21		$ (0.41)
65	**Percent of Net Profit**	14.2%		-0.9%
66				
67	**Desired Rate of Profit**	6%		
68	**Results**	OK		Operating Loss
69				
70				

Figure 15-3: Find a comfortable sale price to avoid too great a loss. Temper this with what profits may have already returned the overhead and desired profit.

Exercise 4

Copy the Overhead Rate to Results labels and formulas and the Markdown columns to the MARKUPCOST range. Edit them as needed and then do the work to make the BUYSELL or all the spreadsheets in this section into templates.

Application Exercises

For the following situations, find the selling price and a reasonable sale price.

1. Caldwell Furniture paid a wholesale distributor $750 apiece for 3 dining room suites. Caldwell must obtain a markup based on selling price of 33.3% to cover their overhead (rate of overhead is 12.7%). They want a 16% rate of profit. What would be the lowest price they could charge before suffering an operating loss based on cost? What markdown rate would break even? (***Hint:*** *Experiment with various Markdown rates to 2 decimal places. Be aware that you may not be able to arrive at the exact answer.*)

2. During a pre-inventory clearance, a $90 pair of boots was advertised at a 40% reduction. If the boots cost $48 originally and overhead is 25% of cost, what was the profit or loss on the sale boots?

3. A 33.3% discount was offered on a tape deck that regularly sold for $96. The tape deck cost $50 and operating expenses were 30% of cost. What was the net profit or loss on the tape deck?

4. A $35 shirt was reduced by 20%. The shirt cost the merchant $22 and expenses are 30% of sales. How much profit or loss was made on the sale?

5. A jewelry store paid $25 for a silver plated water goblet. Their expenses were 15% of the regular selling price and they expect to make a net profit of 17.5% of sales. During a 4th of July sale they offered the goblet at a 45% discount. What was the pre-sale price and what was the net profit or loss?

6. A store paid $30 for a set of aluminum cookware. On regular sales, the overhead is 15% of selling price and markup is set at 25%. At a special sale, the store sold the set at a 20% discount. What was the pre-sale price and the sale price? What was the net profit or loss?

7. Packages of golf balls cost a sports store $105 per dozen. The store marks the balls up to cover expenses of 30% on cost and hopes for a net profit of 12.86%. How much do they charge for a single package of golf balls? What is the lowest rate at which they could sell a package before taking an operating loss? (***Tip:*** *Add the Overhead Rate and the Desired Profit rate to find a single Markup rate.*)

8. A shoe store buys hosiery for $18 per dozen. The marked price of the hose includes a 20% markup on cost to cover overhead and 30% for profit. What do they charge for a single pair of hose? At what rate could they reduce the hose before taking an operating loss? What would this sale price be?

9. Composition notebooks cost a variety store $20 per dozen. The store marks up the notebooks to cover expenses of 40% on cost and a net profit of 20%. What is the markup and the pre-sale price? What is the break-even price?

10. A hardware store buys pliers for $60 a dozen. Expenses at the store run 15% of cost and they expect to make a net profit of 10% on cost. What is the pre-sale price and the break-even price? What is the percent of markdown?

Challenge Problems:

A drugstore paid $2 for each box of color film it purchased. The drugstore sold 120 boxes of the color film for $2.75 each. Another 30 boxes of the film were sold during a special promotion for $1.50. Regular prices at the drugstore include a markup of 20% on cost for overhead. What were the total wholesale cost, total handling cost, and total sales? Was a net profit, operating loss, or gross loss made on the 150 boxes of film?

A farmer brought 115 pounds of organic apples to the farmer's market on Saturday morning. She expects to lose 3 to 5 pound to spoilage, samples, and theft. She also estimates her overhead costs to be $74.75. She hopes to make a 65% profit on the crop early in the day and 40% profit late in the day. She is willing to offer small bargains in the early part of the day and greater ones later to avoid having to bring the apples back home. What is the overhead rate per pound? What are her high and low prices early in the day and late in the day? At what price per pound would she experience an operating loss? An absolute loss? How many pounds must she sell to break even?

V. Other Expenses

16. Payroll

There is no single book or publication that could possibly prepare its readers for all the ins and outs of payroll. This chapter doesn't expect to do so either. In fact, if you have been working through the sections and chapters of this text thus far, you probably have the skills needed to create your own payroll spreadsheet template.

Employees contract with their employers to work either a set number of hours per week or to produce a set product or number of products in a set amount of time. They agree to do so at a set rate. The employee's gross wages are then a simple matter of multiplying the time or production times the pay rate. Salespeople get paid by commission, a percentage of their sales. What the employee takes home, however, is the gross wages minus any deductions for Federal Income Taxes, Social Security, Medicare, retirement funds, and other deductions.

Objectives

By working through and completing the exercises below, you will be able to:

- Include the computation of payroll based on an annual salary in a payroll template.
- Include the computation of payroll based on commission in a payroll template.
- Include the computation of payroll based on hourly work in a payroll template.
- Include the computation of payroll based on production in a payroll template.
- Include the computation of deductions in a payroll template.

Payroll Based on Salary

Salaried employees earn an annual rate of pay and agree to work for that amount. If salaried employees work over the legal standard of 40 hours per week, they may or may not be entitled to time-and-a-half overtime pay. If so they are paid at an equivalent hourly pay rate.

In our example, we will assume the salaried workers do not get overtime. The variation will be in the pattern by which the workers get paid. They usually do not get a single yearly sum once a year. You must prepare for Monthly (12 pay periods per year), Semi-Monthly (twice monthly or 24 per year), and Bi-Weekly (every two weeks or 26 per year) pay periods. To find the gross wages for a pay period you must first determine the number of pay periods per year. The formula for gross wages based on salary is:

Gross Wages = Annual Salary / No of Pay Periods

Exercise 1

Marvin Jones is an architect. His annual salary is $36,000. He is paid monthly. How much is his gross wages in a single pay period.

1. Open a new spreadsheet and save it as PAYROLL or start a worksheet in an EXPENSE spreadsheet.
2. Enter the title **Payroll Based on Salary**.
3. Skip a row and enter the column labels: **Name, Annual Salary, Pay Periods, Gross Wages.**
4. Enter the given data from the problem statement above including the number of Pay Periods.
5. Enter the formula to compute gross wages. In our example it is **=B4/C4**.

6. Save the range of cells used PAYSALARY.
7. Format the table as you can see in Figure 16-1 and save the spreadsheet.

A1 | Payroll Based on Salary

	A	B	C	D	E
1	Payroll Based on Salary				
2					
3	Name	Annual Salary	Pay Periods per Year	Gross Wages	
4	Jones, Marvin	$ 36,000.00	12	$ 3,000.00	
5					
6					
7					
8					

Figure 16-1: Salaried employees get paid on a regular basis, but the pay periods vary according to the pay schedule.

Exercise 2

Even though it is unlikely that a business would pay its employees on different pay schedules, expand this payroll based on salary to include the following employees. Sort them by name and enter a SUM function to total the gross wages.

Table 16-1

Name	Annual Salary	Pay Schedule	Gross Wages to Date
Linden, Joyce	$45,000	Bi-Weekly	$20,688.94
Phiden, John	$26,500	Semi-Monthly	$13,204.69
Jurek, Dana	$27,850	Monthly	$27,754.64
Billings, Susan	$57,975	Monthly	$57,776.56

Payroll Based on Commission

The wages of salespeople often depend entirely on their ability to make sales. Many will stabilize their income by taking an advance against their sales and many also get a salary. The business may also require a minimum sales amount before a salesperson begins to earn a commission. Some companies offers a higher rate of commission with higher sales (called a sliding-scale commission).

In our example, we will deal with the simplest scenario of an employee whose income is based entirely on commission. This is called straight commission. The formula for gross wages then becomes:

$$Gross\ Wages = (Sales - Returns\ and\ Allowances) * Commission\ Rate$$

Exercise 3

Edith Reilly receives an 8% commission on her net sales. Find her commission if her gross (original) sales were $5,680 and sales returns and allowances were $180.

1. Open the PAYROLL worksheet.
2. Enter the title **Payroll Based on Commission**.
3. Skip a row and enter the column labels: **Name, Commission Rate, Gross Sales, R&A, Net Sales, Gross Wages.**

4. Enter the given data from the problem statement above.
5. Enter the formula for Net Sales: =Gross Sales - R&A, which in our example is =R28-S28.
6. Enter the formula to find the Gross Wages; in our example this is =Q28*T28.
7. Select the range of cells used and name it PAYCOMMISSION.
8. Format the spreadsheet as you can see in Figure 16-2 and save the spreadsheet.

P25 | Payroll Based on Commission

	P	Q	R	S	T	U	V
25	**Payroll Based on Commission**						
26							
27	**Name**	**Commission Rate**	**Gross Sales**	**R & A**	**Net Sales**	**Gross Wages**	
28	Reilly, Edith	8.0%	$ 5,680.00	$ 180.00	$5,500.00	$ 440.00	
29							
30							
31							
32							

Figure 16-2: Commission is paid based on sales.

Payroll Based on Hourly Work

The hourly rate for a person's labor can vary according to ability, training, experience, and federal minimum-wage laws or union contracts. The Fair Labor Standards Act now prescribes a 40-hour work week of five 8-hour days at a minimum rate of $5.15 per hour. Overtime is set at a standard 1.5 times the worker's regular hourly rate (time-and-a-half). And, though the Act does not require it, "Double time" is often paid for work on Sundays or on holidays.

In some instances, overtime is paid when an employee works more than 8 hours in a day. Alternately, some states require overtime after an employee accumulates 40 hours a week even though on a single day the employee worked over 8 hours.

Since is it very common for workers to work overtime, our example will include this. The formula for gross wages becomes:

*Gross Wages = (Regular Hours * Pay Rate) + (Overtime Hours * (1.5 * Pay Rate)*

Exercise 4

Wilkins-Casey, Inc., has five employees with the following work record for the weekly payroll: D. R. Elliot M-8, Tu-8, W-10, Th-9, F-10, S-4 at $7.50 per hour; Ray Farmer M-10, Tu-8, W-9, Th-7, F-7, S-4 at $9.90 per hour; Troy Givens M-8, Tu-8, W-8, Th-8, F-8, S-0 at $10.13 per hour; Tony Harper M-5, Tu-8, W-6, Th-7, F-8, S-4 at $9.75 per hour; Marion Ivey M-9, Tu-8, W-6, Th-8, F-9, S-4 at $5.80 per hour. Wilkins-Casey, Inc., pays time-and-a-half for overtime, which starts after 40 hours.

1. Open the PAYROLL worksheet.
2. Enter the title **Payroll Based on Hourly Rate**.
3. Skip a row and enter the column labels: **Name, M, T, W, T, F, S, S, Total Hours, Overtime Hours, Regular Rate Per Hour, Overtime Rate, Regular Wages, Overtime Wages, Total Gross Wages**.
4. Enter the given data from the problem statement above.
5. Enter the formula for Total Hours: =SUM(Mon through Sun cells), which in our example is =SUM(Z43:AF43).

6. Enter an IF formula to find the Overtime Hours and not include hours less than forty: =IF(Total hours>40,then Total Hours - 40,otherwise 0). In our example this is =IF(AG43>40,AG43-40,0).
7. Enter the formula for the overtime rate: =Rate per Hour * 1.5. In our example it is =AI43*1.5.
8. Enter the formula to find the wages for regular hours and for overtime hours. In our example, =(AG43-AH43)*AI43, and =AH43*AJ43.
9. Sum regular and overtime pay to find the Total Gross Wages. In our example this is =AK43+AL43.
10. Copy the formulas down to include the other employees.
11. Name the range of cells used as PAYHOURLY.
12. Format your spreadsheet as seen in Figure 16-3 and save the spreadsheet.

Y40 | Payroll Based on Hourly Rate

Payroll Based on Hourly Rate

Name	M	T	W	T	F	S	S	Total Hours	Overtime Hours	Regular Rate Per Hour	Overtime Rate	Regular Wages	Overtime Wages	Total Gross Wages
Elliot, D. R.	8	8	10	9	10	4	0	49	9	$ 5.00	$ 7.50	$ 200.00	$ 67.50	$ 267.50
Farmer, Ray	10	8	9	7	7	4	0	45	5	$ 6.60	$ 9.90	$ 264.00	$ 49.50	$ 313.50
Givens, Troy	8	8	8	8	8	0	0	40	0	$ 10.13	$ 15.20	$ 405.20	$ -	$ 405.20
Harper, Tony	5	8	6	7	8	4	0	38	0	$ 9.75	$ 14.63	$ 370.50	$ -	$ 370.50
Ivey, Marion	9	8	6	8	9	4	0	44	4	$ 9.75	$ 14.63	$ 390.00	$ 58.50	$ 448.50

Figure 16-3: The weekly payroll for Wilkins-Casey, Inc.

Exercise 5

Copy and adjust the PAYHOURLY range to include overtime by day. Use Figure 16-4 to guide you.

X92 | Payroll Based on Hourly Rate

Payroll Based on Hourly Rate

Name		M	T	W	T	F	S	S	Total Hours	Regular Rate Per Hour	Regular Wages	Total Gross Wages
Elliot, D. R.	Regular Time	8	8	8	8	8			40	$ 5.00	$ 200.00	
	Over Time			2	1	2	4	0	9	$ 7.50	$ 67.50	$ 267.50
Farmer, Ray	Regular Time	8	8	8	7	7			38	$ 6.60	$ 250.80	
	Over Time	2		1			4	0	7	$ 9.90	$ 69.30	$ 320.10
Givens, Troy	Regular Time	8	8	8	8	8			40	$ 10.13	$ 405.20	
	Over Time						0	0	0	$ 15.20	$ -	$ 405.20
Harper, Tony	Regular Time	5	8	6	7	8			34	$ 9.75	$ 331.50	
	Over Time						4	0	4	$ 14.63	$ 58.50	$ 390.00
Ivey, Marion	Regular Time	8	8	6	8	8			38	$ 9.75	$ 370.50	
	Over Time	1				1	4	0	6	$ 14.63	$ 87.75	$ 458.25
	Total Gross Wages											$1,841.05

Figure 16-4: Hourly payroll record for overtime paid per day.

Payroll Based on Production

Production-based payroll challenges the spreadsheet with provisions for overtime and reductions for dockings. Overtime for a production worker, one who is paid per piece or by performance of a task, is based on the worker's hourly rate. Since the hourly rate is dependent on production you must make a formula that divides the regular work-hours' production by the number of regular hours. This looks like:

*Regular Earnings = Pieces * Piece rate*

*Hourly rate = (Pieces * Piece Rate) / Regular hours*

*Overtime Earnings = Overtime hours * (1.5 + Hourly Rate)*

In addition to paying for work done, the production-based business also deducts the cost of spoilage caused by the employee. These penalties are called dockings and are charged at a lower rate than the piece rate since some spoilage is to be expected. The worker gets penalized twice for the spoilage: first, the number of spoiled units is subtracted from the worker's gross units to find the net units; second, the spoiled units are multiplied by the docking rate and this amount is subtracted from the worker's gross wages.

Production workers can also often take advantage of premium rates offered after meeting a minimum production. Thus, they will be paid at a higher piece rate after producing a targeted amount of pieces. The final formula for Gross Wages then becomes:

*Bonus = Bonus Pieces * Premium Rate*

*Docking = Spoilage * Spoilage Rate*

Gross Wages = Regular Earnings + Overtime Earnings + Bonus - Docking

Exercise 6

Cynthia Simmons sews zippers into dresses for a garment manufacturer. Her base rate is 75¢ per dress, and a premium rate of 92¢ per dress is paid after she exceeds the weekly production quota of 240 zippers. She is docked 55¢ for each zipper that does not pass inspection. She does not work overtime, that is, Saturday or Sunday. Compute her gross earnings given the information in Table 16-2.

Table 16-2

Gross Production	54	51	55	53	52
Spoilage	2	0	0	3	1

1. Open the PAYROLL worksheet.
2. Go to a new section of the spreadsheet and enter the title **Payroll Based on Production**.
3. Skip a row and enter the column labels starting two columns to the right: **M, T, W, T, F, Total Regular, Pieces above Quota, Piece Rate, Regular Earnings, Quota, Premium Rate, Bonus, Docking Rate, Total Docked, S, S, Hours Overtime, Overtime Rate, Total Overtime, Gross Wages.**
4. On the row below the column headings and to the left, enter the column heading for **Name** starting in line with the title and then adjacent to this enter row labels, **Gross Production** and then below this **Spoilage** and then **Net Production**. (See Figure 16-5).
5. Enter the given data from the problem statement above.
6. Enter the formula for Total Regular production: =SUM(M through F cells), which in our example is **=SUM(BJ75:BN75)**.
7. Enter the formula for total regular spoilage: = SUM(M through F cells), which in our example is =SUM(BJ76:BN76).
8. Enter a formula to subtract Monday's spoilage from Monday's Gross Production. In our example this is **=BJ75-BJ76**.
9. Copy this formula to the right from the M column to the Total Regular column.

10. Find the number of Pieces Above the Quota: = (Total Pieces - Spoilage)-Quota, which in our example is =BO77-BS77.
11. Enter an IF formula for Regular Earnings that controls for the bonus amount: =IF(Pieces above Quota>0,then (Pieces-Above Quota)*Piece Rate,otherwise Pieces*Piece Rate), which in our example is =IF(BP77>0,(BO77-BP77)*BQ77,BO77*BQ77).
12. Enter the formula for the bonus amount: =Pieces above quota * Premium Rate, which in our example is =BP77*BT77.
13. Enter a formula to find the cost of spoilage (Total Docked) including spoilage from Saturday and Sunday: =(Total Regular Spoilage + Overtime Spoilage) * Docking Rate, which in our example is =(BV77+BX76+BY76)*BO76.
14. Enter the formula to find the overtime: =Saturday + Sunday, which in our example is =BX75+BY75.
15. Enter the formula for the overtime rate: =(Regular Earnings/40) * 1.5; in our example it is =(BR77/40)*1.5
16. Enter the formula to find Total Overtime: Overtime Hours * Overtime rate, which in our example is =BZ77*CA77.
17. Enter the formula to find the Total Gross Wages. In our example this is =BR77+BU77-BW77+CB77.
18. Name the range of used cells and three extra rows for expansion, PAYPRODUCT.
19. Format and compare your work to Figure 16-5 and save the spreadsheet.

BH72 | Payroll Based on Production

	BH	BI	BJ	BK	BL	BM	BN	BO	BP	BQ	BR	BS	BT
72	**Payroll Based on**	**Production**											
73													
74			**M**	**T**	**W**	**T**	**F**	**Total Regular**	**Pieces above Quota**	**Piece Rate**	**Regular Earnings**	**Quota**	**Premium Rate**
75	**Name**	**Gross Production**	54	51	55	53	52	265					
76	Simmons, Cynthia	**Spoilage**	2	0	0	3	1	6					
77		**Net Production**	52	51	55	50	51	259	19	$ 0.75	$ 180.00	240	$ 0.92
78													
79													

BH72 | Payroll Based on Production

	BH	BI	BU	BV	BW	BX	BY	BZ	CA	CB	CC
72	**Payroll Based on**	**Production**									
73											
74			**Bonus**	**Docking Rate**	**Total Docked**	**S**	**S**	**Hours Overtime**	**Overtime Rate**	**Total Overtime**	**Gross Wages**
75	**Name**	**Gross Production**				0	0				
76	Simmons, Cynthia	**Spoilage**				0	0				
77		**Net Production**	$ 17.48	$ 0.55	$ 3.30			0	$ 6.75	$ -	$ 194.18
78											
79											

Figure 16-5: Though production-based employment may seem a simple concept, it makes for a complicated payroll computation even with a spreadsheet.

Deductions

Net income or take-home pay is computed by subtracting regular and special deductions from the gross wages. These deductions can be simple percentage calculations based on gross wages or, as in the case of Federal Income Taxes and many retirement funds, can be somewhat if not very complicated. Thus:

$$Net\ Pay = Gross\ Wages - Deductions$$

The three most common deductions are income taxes, OASDI (Old Age Security and Disability Insurance) and Medicare. FICA (Federal Insurance Contributions Act) was formerly the combination of

Medicare and OASDI, and until recently it was limited to 7.65% of wages up to $62,600. Now there is no limit set on Medicare collection and its rate is at 2.45%. OASDI still halts at the $62,600 earned income cutoff point and now is computed at a rate of 6.2%.

Other deductions can be regular, such as payments to retirement funds, health care, and charitable contributions. Some employees may also have liens on their wages forcing them to pay debts directly from their earnings. There are many possible other examples. For simplicity's sake we will deal with Federal Income Taxes (using a 15% estimated rate though taxes are computed at different rates for different income levels), OASDI, and Medicare deductions.

Exercise 7

Compute the OASDI and Medicare for Marvin Jones from Exercise 1. He has gross wages to date of $35,876.78 and must pay approximately 15% of his gross wages in Federal Income Taxes.

1. Open the PAYROLL worksheet.
2. Go to the Salary range.
3. Add the column labels: **Gross Wages to Date, Medicare, OASDI, Federal Income Tax, Other, Other, Total Deductions, Net Pay**.
4. Enter the subtitle **Deductions** one row above the Gross Wages to Date heading.
5. Enter the given data from the problem statement above.
6. Enter the formula for Medicare: =Gross Wages * 2.45%, which in our example is =0.0245*D4.
7. Enter an IF formula for OASDI payments: =IF(AND(Gross wages<=62600,Gross Wages+Gross Wages to Date<=62600),then Gross Wages*0.0245,otherewise IF(Gross Wages+Gross Wages to Date>62600,then (Gross Wages+Gross Wages to Date)-62600)*0.0245,otherwise 0)), which in our example is =IF(AND(E4<=62600, E4+D4<=62600),D4*0.062,IF(D4+E4>62600,((D4+E4)-62600)*0.062,0)).
8. Enter the formula for estimating Federal Income Taxes: Gross Wages * 15%, which in our example is =D4*.15.
9. Sum the Total Deductions. In our example this is =SUM(F4:J4).
10. Enter the formula for computing Net Pay: Gross Wages - Total Deductions, which in our example is =D4-K4.
11. Format the spreadsheet as you can see in Figure 16-6 and save the spreadsheet.

E1	Deductions								
	A	E	F	G	H	I	J	K	L
1	Payroll Based on	Deductions							
2									
3	Name	Gross Wages to Date	Medicare	OASDI	Federal Income Tax	Other	Other	Total Deductions	Net Pay
4	Jones, Marvin	$ 35,876.78	$ 73.50	$ 186.00	$ 450.00	0	0	$ 709.50	$2,290.50
5									
6									
7									
8									

Figure 16-6: Deductions are easily subtracted from Gross Wages but each one can have a myriad of calculations, rules, and regulations to follow.

Exercise 8

Using the Gross Wages figures, complete the deductions for all the employees in all our examples.

Application Exercises

The following exercises may be asking you to innovate and create your own payroll spreadsheets, such as combining salary and commission or controlling for overtime paid on a daily basis.

1. Find the gross earnings for each of the following hourly employees and the total payroll for this pay period. Regular income is based on a 40-hour work week.

The Software Center								
Name	*M*	*T*	*W*	*T*	*F*	*S*	*S*	*Rate*
Conover, Ray	8	6	7	7	9	0	0	$ 5.10
Evans, D.L.	9	8	4	8	8	3	0	$ 4.45
Hotchkiss, S.P.	7	7	8	6	8	3	0	$ 4.80
Myers, Andrew	8	8	8	8	8	0	0	$ 5.15
Overton, Ross	9	8	8	6	6	0	0	$ 4.90
Royer, Robert	8	7	7	8	6	4	0	$ 5.85
Young, Steven	6	7	8	8	7	3	0	$ 6.00

2. Brian Williams is a salaried employee who earns $95,256 and is paid monthly. What is his pay each payroll period?

3. John Edmonds is paid on a biweekly basis and earns $36,348 annually. Find his first-period gross earnings.

4. Arsella Gallagher earns a salary of $63,552 and is paid semimonthly. What is her gross salary for each pay period?

5. Veronia Reed is paid an annual salary of $53,872. She is paid monthly. What is her gross salary for the month of June?

6. Given the following information, compute the weekly gross wages for each employee of the Micro Management Corporation.

Micro Management Corporation									
Name	*M*	*T*	*W*	*T*	*F*	*S*	*S*	*Rate*	*Regular Hours*
Allen, H.	8	9	8	7	10	4	0	$ 9.86	40
Brown, J.	4	6	8	9	9	5	0	$ 4.97	40
Pick, J.	8	8	8	8	8	4	0	$ 6.87	40
Sayer, c.	9	10	8	9	11	9	0	$ 5.82	40
Lovet, L.	8	8	8	8	0	0	0	$ 7.15	32
Stacy, C.	8	8	8	8	8	0	0	$ 8.21	40

7. Compute David Pressley's gross wages based on the following: Overtime at time-and-a-half is paid daily on work in excess of 8 hours and weekly on work in excess of 40 regular hours; double time is paid for all Sunday work. His hours each day were: M-8, T-9, W-7, T-10, F-7, S-5, S-5.

8. Jewel Warner is a salesperson and is paid a salary of $200 plus 3% of all sales. Find her weekly gross income if her sales are $8.000. There are no returns.

9. William Kelly is a real estate salesperson and receives a 6% commission on the sale of a piece of property sold for $130,000. Find his gross income from this sale.

10. Vincent Ores is paid a salary of $20,800 per year plus 8% of sales. He is paid bi-weekly. Calculate his gross income if his new sales for the current pay period are $9,890.

11. Employees are paid \$3.50 per piece for a certain job. In a week's time, Martin King produced a total of 78 pieces. Find his gross earnings for the week.

12. Jason Richards works in the shipping department of a sweater factory. His base rate is 96¢ per dozen sweaters. He receives a premium rate of \$1.25 per dozen when he packs more than 200 dozen sweaters per week. There is no spoilage rate. Determine his wages for the following: M-42, T-38, W-36, T-44, F-43.

13. Elena Thomas cleans crabs at a fish processing plant. Her base rate is 44¢ per pound of crabmeat. She receives a premium rate of 50¢ per pound when she cleans more than 200 pounds per week. Determine her wages for the following production: M-60, T-62, W-68, T-66, F-70.

14. Randy Gordon operates a drill press on a furnace assembly line. He is paid 2¢ per bore on net production, and a premium rate of 2.5¢ per bore in excess of 9,600 per week. His spoilage chargeback rate is 1.75¢ per bore. Compute his gross wages for the following week: M-2,005, 12; T-1,980, 11; W-19, 72, 8; T-1,978, 10; F-1,994, 11.

15. Sarah Hubbard assembles parts in microcomputers. He base rate is 85¢ on her net production up to 100 completed assemblies. She receives an incentive rate of 92¢ per assembly in excess of 100. She is docked 22¢ for each assembly that does not pass inspection. Calculate her wages for the week when she had the following production: M-26, 1; T-34, 2; W-28, 0; T-34, 3; F-29, 0.

16. Irene Gamble earns \$25,220 per year and is paid bi-weekly. Deductions are as follows: Federal Taxes, \$47; Retirement, \$24.95; Health insurance, \$8.445. Find her net income for a single pay period where her gross to date is \$6,305. (Ignore Medicare and OASDI.)

17. For the following salaried employees, compute the net earnings for the current pay period. They are all paid Bi-weekly. (Include Medicare and OASDI.)

Name	Annual Salary	Pay Schedule	Gross Wages to Date	Other
Abrams, Tom	\$22,620	Bi-Weekly	\$16,965.00	\$24.95
Cowgill, Phylis	\$75,500	Semi-Monthly	\$56,625.00	\$156.75
Mason, George	\$80,000	Monthly	\$61,280.00	\$198.33
Sachs, F. J.	\$49,060	Monthly	\$36,795.00	\$57.36

17. Insurance

Insurance is a service that protects a business from the financial consequences of sudden loss. There are many kinds of insurance offered. We will cover three common kinds of insurance: fire insurance, automobile insurance, and life insurance. An insurance spreadsheet uses lookup tables and rates.

Objectives

By working through and completing the exercises below, you will be able to:

- Find the annual premium for fire insurance on a building and its contents using a fire insurance spreadsheet.
- Find the annual premium for automobile insurance using an automobile insurance spreadsheet.
- Find the annual premium for a life insurance policy using a life insurance spreadsheet.

Fire Insurance

Should fire strike the building your business uses, both the building and its contents are at risk. Fire insurance can help defray the cost of recovery from such a disaster. To compute the expense of fire insurance you will need to consult rate tables. These tables take into consideration the area in which the building is located and the "class" of the building. Some buildings and areas are rated more fire-proof than others and so a Class A building in an area ranked number 1 has the lowest premium rate.

Fire insurance rates are quoted by the per $100 of coverage so the annual premium uses the coverage amount (the face value) divided by 100, and multiplies this times the rate. The formula then is:

$$Annual\ Premium = (Face\ Value\ /100) * Rate$$

You will make your spreadsheet similar to the work done to find a mortgage. First you must build the rate tables for building and contents and plug these into the annual premium formula.

Exercise 1

Build the following tables:

Table 17-1: Building Fire Insurance Rates per $100 of Coverage.

Building			
	Building Classification		
Area Rank	A	B	C
1	$ 0.59	$ 0.75	$ 0.94
2	$ 0.60	$ 0.76	$ 0.96
3	$ 0.62	$ 0.78	$ 0.98

Table 17-2: Contents Fire Insurance Rates per $100 of Coverage.

Contents			
	Building Classification		
Area Rank	A	B	C
1	$ 0.64	$ 0.89	$ 1.11
2	$ 0.65	$ 0.91	$ 1.13
3	$ 0.67	$ 0.93	$ 1.16

The building owned and occupied by O'Toole's Hardware is insured for $85,000. Its contents are insured for $50,000. If it is a Class B building located in Area 2, find the annual premium for building and contents.

1. Open a new spreadsheet or worksheet and save it as INSURE.
2. Enter the title **Fire Insurance**.
3. Skip a row and enter the row labels: **Building Coverage, Contents Coverage, Building Classification, Area,** *blank row,* **Building Rate, Building Annual Premium**, *blank row,* **Contents Rate, Contents Annual Premium**, *blank row,* **Total Premium**.
4. Enter the given data from the problem statement above.
5. In the Building table, select the A, B, and C cells and name the range BLGCLASS.
6. In the Building table again, select the 1, 2, and 3 cells and name the range BLGRANK.
7. In the Building table again, select the range from $.59 to $.98 and name it BLGRATE.
8. In the Contents table, select the A, B, and C cells and name the range CONTENTCLASS.
9. In the Contents table again, select the 1, 2, and 3 cells and name the range CONTENTRANK.
10. In the Contents table, select the range from 1 to $1.16 and name it CONTENTRATE.
11. Enter the formula to find the Building Rate: = INDEX(BLGRATE, MATCH(Area, BLGRANK,Exact), MATCH(Building Class, BLGCLASS, Exact)). In our example this is =INDEX(BLGRATE,MATCH(B6,BLGRank,0),MATCH(B5,BLGClass)).
12. Enter the formula for the Building Annual Premium: =(Building Coverage/100) *Building Rate, which in our example is =(B3/100)*B8.
13. Enter the INDEX formula to find the Contents Rate: =INDEX(CONTENTRATE, MATCH(Area, CONTENTRANK, Exact),MATCH(Building Classification, CONTENTCLASS, Exact)), which in our example is **=INDEX(CONTENTRATE, MATCH(B6,CONTENTRANK,0),MATCH(B5,CONTENTCLASS,0))**
14. Enter the formula for the Contents Annual Premium. In our example it is =(Contents Coverage/100)*Contents Rate, which in our example is **=(B4/100)*B11**.
15. Add the Building Annual Premium and the Contents Annual Premium to find the Total Annual Premium. In our example it is **=B9+B12**.
16. Select and name the used cells the INSFIRE range.
17. Format the table as you can see in Figure 17-1 and save the spreadsheet.

A1 | Fire Insurance

	A	B	C	D	E	F	G	H
1	**Fire Insurance**							
2						**Building**		
3	**Building Coverage**	$85,000.00				**Building Classification**		
4	**Contents Coverage**	$50,000.00			**Area Rank**	**A**	**B**	**C**
5	**Building Classification**	B			1	$ 0.59	$ 0.75	$ 0.94
6	**Area**	2			2	$ 0.60	$ 0.76	$ 0.96
7					3	$ 0.62	$ 0.78	$ 0.98
8	**Building Rate**	$ 0.76						
9	**Building Annual Premium**	$ 646.00				**Contents**		
10						**Building Classification**		
11	**Contents Rate**	0.91			**Area Rank**	**A**	**B**	**C**
12	**Contents Annual Premium**	$ 455.00			1	$ 0.64	$ 0.89	$ 1.11
13					2	$ 0.65	$ 0.91	$ 1.13
14	**Total Premium**	$ 1,101.00			3	$ 0.67	$ 0.93	$ 1.16
15								

Figure 17-1: The annual premium for fire insurance must be found for both the building and its contents.

Automobile Insurance

Fire insurance is a familiar expense to businesses and individuals, especially homeowners. Automobile insurance is also familiar to both though the rates for a commercial vehicle are generally higher due to its high level of use. Finding the annual automobile insurance rate also requires two tables. One table serves finding the Bodily Injury rate while the other lists the Property Damage rate. You must prepare for factors of territory, driver class, and the bodily injury coverage ranges of 25/50, 50/100, and 100/300, and the property damage coverage of 10, 25, and 50. For bodily injury, 25/50 means the insurance company will pay up to $25,000 for a single individual's injuries and up to $50,000 per accident. The property coverage comes in straight $10,000, $25,000, and $50,000 limits.

Exercise 2

Create the following tables:

Table 17-3: Bodily Injury rates table.

Bodily Injury Automobile Insurance Table			
	25/50	50/100	100/300
Territory/Class	1	2	3
1A	$ 287	$ 330	$ 378
1B	$ 417	$ 480	$ 551
1C	$ 132	$ 841	$ 965
2A	$ 194	$ 218	$ 248
2B	$ 281	$ 318	$ 363
2C	$ 491	$ 556	$ 633

Table 17-4: Property Damage rates table.

Property Damage Automobile Insurance Table			
Territory/Class	10	25	50
1A	$ 155	$ 178	$ 204
1B	$ 225	$ 258	$ 297
1C	$ 394	$ 453	$ 519
2A	$ 104	$ 118	$ 134
2B	$ 151	$ 172	$ 195
2C	$ 265	$ 300	$ 341

Find the annual premium for an automobile liability insurance policy in which the insured lives in Territory 1, is Class A, and wishes to have 50/100/10 coverage.

1. Create a new area in the INSURE spreadsheet or a new worksheet.
2. Enter the title **Automobile Insurance**.
3. Enter the following row labels: **Territory/Class, Bodily Injury Coverage, Bodily Injury Premium,** *blank row, blank row,* **Property Damage Coverage, Property Damage Annual Premium,** *blank row,* **Total Annual Premium**
4. Enter the given data.
5. In the Bodily Injury table, select the range from entry 1 to entry 3 and name it BICOVERAGE.
6. In the Bodily Injury table, select the range from entry 1A down to entry 2C and name it BITYPE.
7. Again in the Bodily Injury table, select from $287 to $633 and name the range BIRATE.

8. Enter the INDEX function formula to find the Bodily Injury Premium: =INDEX(BIRATE, MATCH(Territory/Class, BITYPE, Exact), MATCH(Bodily Injury Coverage, BICOVERAGE, Exact)), which in our example is =INDEX(BIRATE,MATCH (L42,BITYPE,0),MATCH(L43,BICOVERAGE)).
9. In the Property Damage table, select the range from entry 10 to entry 50 and name it PROPCOVERAGE.
10. Again in the Property Damage table, select from $155 to $341 and name the range PROPRATE.
11. In the Property Damage table, select from entry 1A to entry 2C and name it PROPTYPE.
12. Enter the INDEX function formula to find the Property Damage Annual Premium: =INDEX(PROPRATE,MATCH(Territory/Class,PROPTYPE,Exact),MATCH(Property Damage Coverage,PROPCOVERAGE)) =INDEX(PROPRATE,MATCH(L42, PROPTYPE,0),MATCH(L47,PROPCOVERAGE)).
13. Add the two premiums to find the Total Annual Premium. In our example this is =L44+L48.
14. Select the range of cells used and name them INSAUTO.
15. Format the table as you can see in Figure 17-2 and save the spreadsheet.

K40 | Automobile Insurance

	K	L	M	N	O	P	Q	R	S
40	**Automobile Insurance**								
41					**Bodily Injury Automobile Insurance Table**				
42	**Territory/Class**	1A				25/50	50/100	100/300	
43	**Bodily Injury Coverage**	2			Territory/Class	1	2	3	
44	**Bodily Injury Premium**	$ 330.00			1A	$ 287	$ 330	$ 378	
45					1B	$ 417	$ 480	$ 551	
46					1C	$ 132	$ 841	$ 965	
47	**Property Damage Coverage**	10			2A	$ 194	$ 218	$ 248	
48	**Property Damage Annual Premium**	$ 155.00			2B	$ 281	$ 318	$ 363	
49					2C	$ 491	$ 556	$ 633	
50	**Total Annual Premium**	$ 485.00							
51					**Property Damage Automobile Insurance Table**				
52					Territory/Class	10	25	50	
53					1A	$ 155	$ 178	$ 204	
54					1B	$ 225	$ 258	$ 297	
55					1C	$ 394	$ 453	$ 519	
56					2A	$ 104	$ 118	$ 134	
57					2B	$ 151	$ 172	$ 195	
58					2C	$ 265	$ 300	$ 341	
59									

Figure 17-2: find the Bodily Injury annual premium and the Property Damage annual premium to find how much you will pay for automobile insurance for your territory, driver class, and coverage.

Life Insurance

Occasionally a business finds it advantageous to insure an individual's life, but usually this is something we do for ourselves. Sometimes the business will offer such insurance along with medical insurance for which the employer makes a contribution.

Life insurance comes in two basic varieties: Term insurance and Straight-life. Term insurance is purchased for a certain period of time such as 5, 10, and 20 years. The insured person pays during the term of the policy and, if the person does not die, the policy ends when the term expires and the policy has no cash value.

Straight-life policies gain in value like an annuity. The insured pays into the policy for life or until the insured "cashes" the policy in. It is also available for use as collateral for loans. Straight-life insurance also costs more in the short run than does term insurance.

Life insurance is bought at rates per $1,000 of coverage. The spreadsheet uses two tables, one for each kind of coverage. The age and sex of the insured person is taken into consideration, with older men paying the highest rates. The formula for the annual premium for either term or straight-life is:

$$\textit{Annual Premium} = (\textit{Face Value} / 1000) * \textit{Rate}$$

We will also use a VLOOKUP function instead of the INDEX function since the tables only have two rows. The VLOOKUP function will return a value in an array corresponding to a lookup value. You will pass an age from the Age column in the tables and then use IF functions to pick the table and then either 5-year Term insurance or Straight life insurance.

Exercise 3

Create the following tables, one above another.

Table 17-5: Term and Straight-life insurance for Males.

Life Insurance Premium Table		
Male		
Age	5 yr Term	Straight
20	$ 5.26	$ 12.03
30	$ 5.50	$ 15.25
40	$ 7.86	$ 21.41
50	$ 15.46	$ 32.99
60	$ 31.40	$ 54.67

Table 17-6: Term and Straight-life insurance for Females.

Female		
Age	5 yr Term	Straight
20	$ 4.96	$ 11.06
30	$ 5.16	$ 13.70
40	$ 6.46	$ 17.97
50	$ 10.70	$ 26.37
60	$ 17.74	$ 42.30

Find the annual premium of an insurance policy with a face value of $15,000 for a 30-year-old male for both a 5-year term policy and a straight-life policy.

1. Start a new section in the INSURE spreadsheet or start a new worksheet.
2. Create the title cell **Life Insurance**.
3. Add the following row labels: **Face Value, Sex, Age, Policy Type**, *blank row*, **Rate**, *blank row*, **Annual Premium**
4. Enter the given data from the problem statement using M to stand for Male and Straight for the Policy Type.
5. In the Male table, select from $20 to $54.67 and name the range MALERATE.

6. In the Female table, select from $4.96 to $42.30 and name the range FEMALERATE.
7. Enter a VLOOKUP function with embedded IF functions to find the premium rate for this policy taking into consideration the age and sex of the insured person: =VLOOKUP(Age, IF(Sex="M",then MALERATE,otherwise FEMALERATE),IF(Policy Type="Straight",then column 3,otherwise column 2)), which in our example is =VLOOKUP(W77, IF(W76="M",MaleRate,FemaleRate),IF(W78="Straight",3,2)).
8. Enter the formula for computing the Annual Premium. In our example it is =(W75/1000)*W80.
9. Record the amount for straight-life and substitute Term for the Policy Type to find the alternative policy price.
10. Select the used cells and give this range the INSLIFE range.
11. Format your spreadsheet as seen in Figure 17-3 and save the spreadsheet.

V73 | Life Insurance

	V	W	X	Y	Z	AA	AB
73	Life Insurance						
74					Life Insurance Premium Table		
75	Face Value	$25,000.00					
76	Sex	M			Male		
77	Age	35		Age	5-Yr Term	Straight	
78	Policy Type	Straight		20	$ 5.26	$ 12.03	
79				30	$ 5.50	$ 15.25	
80	Rate	$ 15.25		40	$ 7.86	$ 21.41	
81				50	$ 15.46	$ 32.99	
82	Annual Premium	$ 381.25		60	$ 31.40	$ 54.67	
83							
84					Female		
85				Age	5-Yr Term	Straight	
86				20	$ 4.96	$ 11.06	
87				30	$ 5.16	$ 13.70	
88				40	$ 6.46	$ 17.97	
89				50	$ 10.70	$ 26.37	
90				60	$ 17.74	$ 42.30	
91							

Figure 17-3: A Life Insurance spreadsheet is simpler than the others though it does depend on a fairly complicated VLOOKUP formula.

Application Exercises

For the following exercises, you may be asked to innovate to handle situations common to the insurance scenario but not specifically covered in the examples given. However, do use the tables in this chapter for the necessary rates.

1. Audie Cavanaugh owns a Class B office building located in Area 2. What is the annual fire insurance premium if the building is insured for $350,000 and the contents are insured for $100,000?
2. Find the annual fire insurance premium on a Class A building located in Area 3 if the building is insured for $120,000 and its contents are insured for $75,000.
3. A Class C building and its contents are located in Area 1 and are insured for $150,000 and $68,000 respectively. What is the total annual insurance premium? If a 2% charge is added to the

annual premium when payments are made semiannually, how much would semiannual payments be?

4. Alice Murillo owns a Class A building located in Area 1. The building is insured for $200,000 and the contents are insured for $85,000. A 2% charge is added to the annual premium because she pays the insurance quarterly. How much is her quarterly payment?

5. Find the annual premium for an automobile liability insurance policy if the insured lives in Territory 2 and is classified as a Class C driver. The policy contains 25/50/10 coverage.

6. Compare the annual premium on a 50/100/10 policy for a Class C driver in Territory 1 to a policy with the same coverage for a Class C driver in Territory 2.

7. What are the monthly payments on an automobile liability insurance policy for a Class B driver in Territory 1 with 50/100/25 coverage? Assume no additional fee is required for the monthly payment option.

8. Find the annual premium for a life insurance policy with a face value of $45,000 for a 22-year-old female for a 5-year term policy.

9. What are the quarterly payments on a $100,000 straight-life insurance policy for a 31-year-old male?

10. Compare a 5-year term policy for $75,000 for a 40-year-old male to the same policy for a 49-year-old female.

11. What are the monthly payments on a $50,000 straight-life insurance policy for a 57-year-old male?

12. Find the annual premium for a straight-life insurance policy with a face value of $45,000 for a 28-year-old female.

13. Find the building, contents and total annual fire insurance premium for each of the following:

	Area Rank	Class	Building	Contents
1	3	A	$ 72,000.00	$26,000.00
2	1	C	$ 38,000.00	$21,000.00
3	2	B	$116,000.00	$41,700.00
4	2	A	$ 78,500.00	$32,300.00
5	3	C	$105,000.00	$63,500.00
6	1	B	$258,000.00	$79,000.00

14. Find the automobile annual insurance premium for the following:

	Territory	Driver Class	Coverage
1	1	B	25/50/10
2	2	A	50/100/10
3	1	C	100/300/25
4	2	B	50/100/25
5	2	C	50/100/10
6	1	A	100/300/50

15. Find the life insurance premium for each of the following:

	Sex	Age	Policy Type	Face Value
1	Male	20	5-year term	$ 30,000.00
2	Male	30	5-year term	$ 90,000.00
3	Female	20	Straight-life	$ 60,000.00
4	Male	50	Straight-life	$100,000.00
5	Female	60	Straight-life	$ 50,000.00
6	Male	20	Straight-life	$ 40,000.00
7	Female	30	Straight-life	$ 80,000.00
8	Male	40	5-year term	$100,000.00

16. Find the semiannual premium for a 50/100/25 automobile liability insurance policy for a Class A driver in Territory 2 if there is no additional charge for semiannual payment.

17. Find the quarterly payments on a 5-year term life insurance policy for $95,000 on a 40-year-old female.

18. Find the monthly payments on a 5-year term insurance policy for $75,000 for a 30-year-old male.

19. How much are the quarterly premiums on a 50/100/25 automobile liability insurance policy for a Class B driver in Territory 1 if a 2% charge is added to annual premiums that are paid quarterly?

20. Find the face value of a fire-protection policy on a Class A building in area 1 worth $87,500 if it is insured for 75% of its market value.

18. Depreciation

In Chapter 3, you subtracted depreciation from the assets of businesses. Depreciation is a systematic way to recover the cost of an asset over its useful life.

Federal Income Tax offers one way of recording deprecation costs using the MACRS (Modified Accelerated Cost Recovery System) method for which the government provides tables and equipment categories by which to determine depreciation. The government method is only one way to account for depreciation. What you will cover in this chapter will be the more traditional methods: straight-line method, Declining-Balance method, sum-of-the-year's-digits method, and the units-of-production method.

Objectives

By working through and completing the exercises below, you will be able to:

- Prepare a depreciation schedule template using the straight-line method.
- Prepare a depreciation schedule template using the Declining-Balance method.
- Prepare a depreciation schedule template using the Sum-of-the-Year's-Digits method.
- Prepare a depreciation schedule template using the units-of-production method.

Straight-Line Method of Depreciation

The straight-line method of depreciation is the simplest method for computing depreciation and is acceptable for tax purposes in almost all cases. In this method the total allowable depreciation is divided by the asset's useful life. The same amount of depreciation is charged each year. To do so you must first determine the Residual Value of an asset. The residual value is the expected value of the asset at the end of its useful life. It is the value of the asset as a trade-in, or its resale value, or its scrap value. The difference between an asset's cost and its residual value is its depreciable value. You can find the amount to depreciate an asset each year of its useful life (its annual depreciation) by dividing the depreciable value by the number of years in its useful life. You must also keep track of the amount of depreciation from year to year by keeping a running balance for the accumulated depreciation. The main formulas you will use in setting up a depreciation schedule are:

$$\textit{Depreciable Value} = \textit{Cost -Residual Value}$$

$$\textit{Annual Depreciation} = \textit{Depreciable Value/Useful Life in Years}$$

Exercise 1

A machine that manufactures keys is purchased for $5,100. It is expected to last for 5 years and have a trade-in value of $900. Prepare a depreciation schedule by the straight-line method.

1. Open a new spreadsheet and save it as DEPREC.
2. Enter the title **Straight-Line Method**.
3. Skip a row and enter the row labels: **Original Cost, Residual Value**, *blank row,* **Depreciable Value**, *blank row,* **Life in Years**, *blank row,* **Annual Depreciation.**
4. Enter the given data from the problem statement above.
5. Name the Original Cost value's cell the SLCOST range.
6. Enter the formula for depreciable value. In our example it is **=B3-B4**.

7. Enter the formula for the annual depreciation. In our example it is **=B6/B8**. Name this cell the SLANNUALDEP range.
8. In a new section to the right and below this enter the title and subtitle Depreciation Schedule and Straight-Line Method.
9. Enter the column headings of **Year, Book Value (End of Year), Annual Depreciation, Accumulated Depreciation**
10. Enter a series of numbers from 0 to 5 for the Year column.
11. Reference the Original Cost cell for the first Book Value (cell E5): =SLCOST.
12. Enter the reference to the Annual Depreciation for each of the Annual Depreciation cells from years 1 through 5 (not for year 0): =SLANNUALDEP.
13. Enter the formula for the second year's book value (cell E6): First year's book value - Annual Depreciation, which in our example is =E5-SLANNUALDEP.
14. Copy this formula down to the fifth year.
15. Enter the formula for the running balance for Accumulated Depreciation (cell G6): Previous Accumulated + Annual Depreciation for Year 1, which in our example is =G5+F6.
16. Copy this formula down to the fifth year.
17. Select the range of used cells and name it DEPSTRAIGHT
18. Format the table as you can see in Figure 18-1 and save the spreadsheet.

A1 | Straight-Line Method

	A	B	C	D	E	F	G	H
1	**Straight-Line Method**							
2					**Depreciation Schedule**			
3	**Original Cost**	$ 5,100			Straight-Line Method			
4	**Residual Value**	$ 900		Year	Book Value (End of Year)	Annual Depreciation	Accumulated Depreciation	
5				0	$ 5,100			
6	**Depreciable Value**	$ 4,200		1	$ 4,260	$ 840	$ 840	
7				2	$ 3,420	$ 840	$ 1,680	
8	**Life in Years**	5		3	$ 2,580	$ 840	$ 2,520	
9				4	$ 1,740	$ 840	$ 3,360	
10	**Annual Depreciation**	$ 840		5	$ 900	$ 840	$ 4,200	
11								

Figure 18-1: Using the straight-line method of depreciation, the final book value equals the residual value and the total accumulated depreciation equals the depreciable value.

Declining-Balance Method of Depreciation

The Declining-Balance method gives the greatest annual depreciation during the early years of an asset's life. This often reflects the truer cost of the asset since most things depreciate more in the early part of their lives. Using it you first find the Declining-Balance rate, which is twice the reciprocal of the useful life in years. Then you multiply this rate times each succeeding book value:

*Depreciation Rate = 2 * (1 / Useful life in years)*

*Annual Depreciation = Rate * Previous year's book value*

This method will never use up all the depreciable amount since the rate will always find yet another portion of the book value. You must enter an adjustment formula to depreciate the final amount and then, to use the template in the future, you will need to insert rows or delete rows above this final row to adjust for varying useful life-spans:

$$\textit{Final Depreciation} = \textit{Second-to-last Book Value} - \textit{Residual Value}$$

Exercise 2

A new vending machine that cost $4,800 has an expected residual value of $600 and a life of 5 years. Prepare a depreciation schedule by the Declining-Balance method.

1. Copy the workspace and depreciation schedule from the straight-line method worksheet to a new section of the DEPREC spreadsheet or to a new worksheet.
2. Edit the title in the work area to read **Declining-Balance Method**, and name the cell DEPDECLINING. Change the subtitle in the schedule to **Declining-Balance Method**.
3. Enter the given data from the problem statement above.
4. Change the Annual Depreciation row label to **Depreciation Rate**.
5. Enter the formula for the Depreciation Rate. In our example it is =2*(1/K37).
6. Name the Original Cost amount's cell the DBCOST and the Depreciation Rate the DEPRATE.
7. Change the 0 year Book Value to =DBCOST.
8. Enter the formula for the first year's annual Depreciation. In our example, =DEPRATE * DBCOST.
9. For the first year's Book Value, enter a running balance formula to deduct the annual depreciation. =0 year's Book Value - 1st year's annual deprecation, which in our example is =N34-O35.
10. Copy the formulas from steps 8 and 9 down to the fifth year.
11. For the last year of deprecation, enter the adjustment formula. In our example it is =N38-K33.
12. Select the range of used cells and name it DEPDECLINEBAL.
13. Format the table as you can see in Figure 18-2 and save the spreadsheet.

O39 =N38-K33

	J	K	L	M	N	O	P	Q
30	Declining-Balance Method							
31					Depreciation Schedule			
32	Original Cost	$ 4,800			Declining-Balance Method			
33	Residual Value	$ 600		Year	Book Value (End of Year)	Annual Depreciation	Accumulated Depreciation	
34				0	$ 4,800			
35	Depreciable Value	$ 4,200		1	$ 2,880	$ 1,920	$ 1,920	
36				2	$ 1,728	$ 1,152	$ 3,072	
37	Life in years	5		3	$ 1,037	$ 691	$ 3,763	
38				4	$ 622	$ 415	$ 4,178	
39	Depreciation Rate	40%		5	$ 600	$ 22	$ 4,200	
40								

Figure 18-2: The final year of a Declining-Balance depreciation schedule uses a simple subtraction formula to adjust for the final share of the depreciation.

Sum-of-the-Year's-Digits Method

The Sum-of-the-Year's-Digits method may remind you of the work you did to find the Rule of 78 value. Here, to find the Sum-of-the-Year's-Digits you add the digits that account for each year of the asset's life. So, if an asset's life is five years you must add 1 + 2 + 3 + 4 + 5 = 15. Since this process would be cumbersome, you will use the formula:

$$\textit{Sum-of-the-Year's-Digits} = (\textit{Years}(\textit{Years} + 1)) / 2$$

Using the Sum-of-the-Year's-Digits method, the largest amount of depreciation is taken off in the first year and the least amount is taken out in the final year. To compute the annual depreciation you use the Sum-of-the-Year's-Digits figure as the denominator of a series of fractions. The formula multiplies this depreciable value times the fractions each year with the first year using the number of years as the numerator. The formula you will use is:

*Annual Deprecation(Life in Years-Year Number)/Sum-of-the-Year's-Digits)*Depreciable Value*

Exercise 3

Prepare a depreciation schedule for the information found in *Exercise 1*. Use the Sum-of-the-Year's-Digits method.

1. Open the DEPREC spreadsheet and copy the whole of the Declining-Balance work to a new section or to a new worksheet.

2. Change the title of the work area and subtitle for the depreciation schedule to **Sum-of-the-Year's-Digits Method**.

3. Replace the Depreciation Rate row label with the **Sum-of-the-Year's-Digits** row label.

4. Enter the given data from *Exercise 1*.

5. Enter the formula for the Sum-of-the-Year's-Digits. In our example it is =(S66*(S66+1))/2

6. Enter the formula for the first year's Annual Deprecation. In our example it is =(($S$66-U63)/$S$68)*$S$64.

7. Copy the step 6 formula down to the last year.

8. Name the range of used cells to DEPSUMYEARS.

9. Format your spreadsheet as seen in Figure 18-3 and save the spreadsheet.

R59 | Sum-of-the-Year's-Digits Method

	R	S	T	U	V	W	X	Y
59	Sum-of-the-Year's-Digits Method							
60					Depreciation Schedule			
61	Original Cost	$ 5,100			Sum-of-the-Year's-Digits Balance Method			
62	Residual Value	$ 900		Year	Book Value (End of Year)	Annual Depreciation	Accumulated Depreciation	
63				0	$ 5,100			
64	Depreciable Value	$ 4,200		1	$ 3,700	$ 1,400	$ 1,400	
65				2	$ 2,580	$ 1,120	$ 2,520	
66	Life in years	5		3	$ 1,740	$ 840	$ 3,360	
67				4	$ 1,180	$ 560	$ 3,920	
68	Sum-of-the-Year's- Digits	15		5	$ 900	$ 280	$ 4,200	
69								

Figure 18-3: The Sum-of-the-Year's-Digits depreciation method, like the Declining-Balance method, takes the greater portion of value off the asset in its early years.

Units-of-Production Method

What more straight-forward method of allocating the cost of an asset over its useful life then by its use? Consider the key making machine in Exercise 1, every key it makes wears on the machine. You can depreciate by units of production by dividing the depreciable value by the total expected units produced in the machine's useful life. You can then use this rate multiplied times the actual units used per year to find the annual depreciation:

Unit rate = Depreciable Value / Total Units

*Annual Depreciation = Annual Units * Unit Rate*

Exercise 4

Use the units-of-production method to determine annual depreciation for the key machine in Exercise 1. The number of units produced annually is given in Table 18-1. The machine is expected to produce 560,000 keys during its lifetime.

Table 18-1

Year	Annual Units
1	120,000
2	160,000
3.	40,000
4.	80,000
5.	160,000

1. Open the DEPREC spreadsheet and copy the work area and depreciation schedule from the Sum-of-the-Year's-Digits area to a new area or a new worksheet.
2. Change the title of the work area and the subtitle of the schedule to **Units-of-Production Method**.
3. Change the Life in Years row label to **Expected Production** and the Sum-of-the-Year's-Digits row label to **Unit Rate.**
4. Fill in the given data and name the Cost the UNITSCOST range and the Unit Rate the UNITRATE range.
5. Enter the formula for the unit rate. In our example it is =AA88/AA90.
6. Insert a column between the Year and Book Value columns and give it the column heading **Annual Units**.
7. Fill in the annual unit amounts from Table 18-1 above.
8. For the zero year, enter the Book Value referencing the UNITSCOST range: =UNITSCOST.
9. For the first year's annual depreciation (cell AF88), enter the annual depreciation formula. In our example it is =AD88*UNITRATE.
10. Copy this formula down to the last year.
11. Name the used cells the DEPUNITS range.
12. Compare your work to Figure 18-4, format it appropriately, and save the spreadsheet.

DepUnitsProd | Units-of-Production Method

	Z	AA	AB	AC	AD	AE	AF	AG	AH
83	**Units-of-Production**	**Method**							
84						**Depreciation Schedule**			
85	**Original Cost**	$ 5,100				Units-of-Production Method			
86	**Residual Value**	$ 900		**Year**	**Annual Units**	**Book Value (End of Year)**	**Annual Depreciation**	**Accumulated Depreciation**	
87				0		$ 5,100			
88	**Depreciable Value**	$ 4,200		1	120,000	$ 4,200	$ 900	$ 900	
89				2	160,000	$ 3,000	$ 1,200	$ 2,100	
90	**Expected Production**	560,000		3	40,000	$ 2,700	$ 300	$ 2,400	
91				4	80,000	$ 2,100	$ 600	$ 3,000	
92	**Unit Rate**	0.0075		5	160,000	$ 900	$ 1,200	$ 4,200	
93									

Figure 18-4: The units-of-production method accurately accounts for a machine's usage per year.

Application Exercises

1. Davis Auto Shop purchased shop equipment costing $23,000 with a salvage value of $3,000. The useful life was estimated to be 8 years. Construct a depreciation schedule using the straight-line method.

2. Chambers Advertising Co. purchased a computer for $1,450. The company plans to use the computer for 4 years. Construct a depreciation schedule using the straight-line method and assume the trade-in value will be $350.

3. Gold's Department Store purchased new display counters at a cost of $25,000. After 10 years' use, the counters should be worth about $2,500. How much depreciation can be claimed each year by the straight-line method?

4. Determine the annual depreciation using the straight-line method for office desks and chairs costing $5,200, with a salvage value of $500 and a useful life of 10 years.

5. The Oasis Restaurant purchased furnishings for $42,000. The restaurant plans to replace these furnishings in 5 years, at which time the salvage value should be $8,000. Determine the annual depreciation by the straight-line method.

6. The Children's Clinic paid $6,300 for its waiting room furnishings. The furniture will be worth $800 after 5 years of use. Using the straight-line method, determine the annual depreciation for these furnishings.

7. Meg's Craft Shop purchased display cases for $7,000. Meg estimates the life to be 4 years at which time the salvage value will be zero. Use the double Declining-Balance method and construct a depreciation schedule.

8. New personal computers cost an accounting firm $10,000. For financial reporting, the firm has chosen the Declining-Balance method for depreciating these computers. The life is estimated to be 5 years and the trade-in value at that time will be $500. Construct a depreciation schedule.

9. Murphy Appliance Co. uses the Declining-Balance method on shop equipment that cost $35,000. The useful life of the equipment is 10 years. Determine the annual depreciation for the first 4 years only.

10. TV station XYZ paid $7,290 for a new video camera, which should remain servicable for 6 years, at which time the residual value should be $400. Make a depreciation schedule by the Declining-Balance method.

11. A high-speed copier that cost $8,000 will become obsolete in 4 years having a residual value of zero. Determine the yearly depreciation by the Declining-Balance method and construct a depreciation schedule.

12. A photo-processing machine cost Color Corp. $5,000. The machine will be in use for 5 years and have a residual value of $600. Make a depreciation schedule using the Declining-Balance method.

13. Use the Sum-of-the-Year's-Digits method to calculate the annual depreciation for each of the following assets:

	Cost	Scrap Value	Years
1	$ 8,900.00	$ 500.00	6
2	$ 15,200.00	$ 1,200.00	7
3	$ 1,900.00	$ 250.00	10
4	$ 5,300.00	$ 800.00	5
5	$ 7,900.00	$ 1,500.00	4
6	$ 12,000.00	$ 3,000.00	5
7	$ 3,200.00	$ 400.00	7
8	$ 18,000.00	$ 3,600.00	4
9	$ 8,300.00	$ 1,100.00	8
10	$ 3,300.00	$ 150.00	9

14. Use the units-of-production method to compute the unit rate and the annual depreciation for each machine below. In each case assume that the machine is expected to produce 300,000 units during its useful life.

	Cost	Scrap Value	Useful Life	Annual Production
1	$ 2,200	$ 700	5 years	80,000, 60,000, 40,000, 20,000, 100,000
2	$ 4,800	$ 300	4 years	60,000, 120,000, 80,000, 40,000
3	$ 6,300	$ 300	4 years	100,000, 75,000, 80,000, 45,000
4	$ 9,300	$ 900	5 years	25,000, 50,000, 70,000, 95,000, 60,000
5	$ 3,150	$ 750	4 years	40,000, 50,000, 115,000, 95,000
6	$ 5,400	$ 600	6 years	25,000, 40,000, 50,000, 55,000, 60,000, 70,000

Appendix I: Excel Menus and Commands

The following table lists the menus and their commands for Excel 4.0 for Windows. Not all the commands are included. Ones not included are those that have no relevance to this publication and which are not expected to be of immediate use to a new spreadsheet user.

Tip: ***You can discover many keyboard shortcuts by studying the menu and command entries next to which you can see the "hotkeys" for many commands. Other shortcut keyboard methods can be found in the Excel help file under the Keyboard Shortcuts topic.***

Menu	Command	Description
File	New	Opens a new Workbook with four blank worksheets.
	Open	Presents a File Open dialog box, which includes a File Find... button.
	Close	Closes the current Workbook.
	Save	Writes the current changes to the current workbook to disk. If the Workbook has not yet been saved, it presents the Save As dialog box.
	Save As...	Presents the Save As dialog box letting you save the workbook with a new name and in a different directory if desired.
	Save Workspace...	Presents the Save Workspace dialog box, which allows you to save all the open workbooks and their status (how they appear in the application) as a single .xlw file from which you can open all the workbooks at once.
	Find File...	Presents the File Find dialog box from which you can define search criteria for locating and opening a file.
	Summary Info...	Presents the Summary Info dialog box, which lets you record the title, author, and other information about the file.
	Page Setup...	Presents the Page Setup dialog box with its four tabs: Page, Margins, Header/Footer, and Sheet. Use the Page tab to define the paper and its orientation. Use the Margins tab to define the white space on a printed page. Use the Header/Footer tab to set the top and bottom of every page. Use the Sheet tab to define the print order.
	Print Preview...	Presents a WYSIWYG view of the current worksheet as it will print, which includes page setup access.
	Print...	Presents the Print dialog box from which you can specify what pages and how many pages to print as well as access page setup and print preview.
	1, 2, 3, Files	Select and open spreadsheets from this list of previously opened files.
	Exit	Closes the Excel program.
Edit	Undo/Redo	Removes the effects of the most recent command or, if the most recent command was Undo, reestablishes the effects of the Undo command.
	Repeat	Repeats the most recent command.
	Cut	Removes a cell's entry and puts it on the Windows Clipboard.
	Copy	Puts a cell's entry onto the Clipboard.
	Paste	Places the contents of the Windows Clipboard into the selected cell.
	Paste Special...	Presents the Paste Special dialog box to control the paste operation. For instance, select the Value paste option to only place the results of a formula or choose the Transpose option to paste a row of labels across the top of columns.
	Fill	Fill the selected range with values. This command presents another command list to fill Down, Right, Up, Left, Across worksheets..., create a Series..., or to Justify the contents of the selected range. The Across Copies dialog box lets you copy the selected range to other worksheets in the workbook (select All, Contents, and Formats to copy all the data, just its contents or just its formatting). Use the Series dialog box to create a pattern down a column or across a row such as 2, 4, 6, ... or Mon., Tue., Wed., ...
	Clear	Presents another command menu to let you clear All the information in a cell, just

Menu	Command	Description
		the Format, just the Contents, or just the Notes.
	Delete...	With a row or column selected, removes the row or column from the spreadsheet. With a cell or range of cells selected, it presents a Delete dialog box to control how adjacent cells will replace the removed cells.
	Delete Sheet	Removes the current worksheet from the workbook.
	Move or Copy Sheet...	Presents a dialog box that lets you rearrange the position of the current worksheet before or behind other worksheets in a workbook.
	Find...	Presents the Search dialog box from which you can enter text and values for Excel to find and select for you.
	Replace...	Presents the Replace dialog box from which you can enter text and values for Excel to find and replace for you.
	Go To...	Presents the Go To dialog box from which you can select named ranges or you can enter cell references for Excel to select and bring into view on your screen.
View	Formula Bar	Toggles the main window to showing or not showing an area above the worksheet columns that displays the contents of a cell (the cell may show the results while the bar shows the actual contents of a formula). The bar also holds buttons that quickly place or help you create formulas.
	Status Bar	Toggles the main window to showing or not showing the Status bar at the bottom of the window. This bar tells you the conditions under which you are operating the program. For example, if OVR shows on this bar, in text entry you will type over the existing text rather than add to it.
	Toolbars...	Presents a Toolbar dialog box from which you can choose to show or not show various toolbars below the menu or at the bottom of the screen and lets you access tools for modifying toolbars (add or remove buttons).
	Full Screen	Toggles the program window to show only the worksheet and the main menu for a wider view of your work.
	Zoom...	Presents the Zoom dialog box from which you can choose to increase or reduce the size of the view (not the actual size of cells) so that you can see the contents more clearly or see more of the worksheet at one time on screen.
Insert	Cells...	Presents the Insert Cells dialog box from which you can control how adjacent cells will move when you insert a cell or range of cells into the worksheet.
	Rows	Inserts the same number of rows as are rows selected above the selected row(s). The new rows inherit the qualities of the row above them.
	Columns	With columns selected, inserts to the left the same number of columns. The new columns inherit the qualities of the column to their immediate left.
	Worksheet	Adds a worksheet to the workbook in front of or above the current worksheet.
	Chart	Presents a submenu to command the use of a Chart Wizard tool to add a chart (graph) On This Sheet or On New Sheet using the data from the selected cells. After selecting a command, click and drag to define where you want the chart and to open the Chart Wizard.
	Page Break	Defines a page break after the selected cell(s) to control where the page ends and a new page begins.
	Function...	Presents the Function Wizard from which you can select a function and be guided in completing your formula.
	Name	Presents a submenu to Define..., Paste..., Create..., Apply... a name to the selected cell or range. Each opens the Named Ranges dialog box from which you can create (Define) or select named ranges.
	Note...	Presents a dialog box for the composition and storage of a note for a cell or range of cells.
Format	Cells...	Presents the Format Cell dialog box with its tabs for defining the type of Number used in a cell, the content's Alignment, the Font used, the Borders used, the Patterns used, or to Protect or unprotect the selected cells. (Protect the worksheet using the Tools, Protection command.)

Menu	Command	Description
	Row	Opens a submenu to control the size of the row cells. The Height… option opens a dialog box in which you can define the height. The AutoFit command adjusts the height to the size of the tallest entry. The Hide and Unhide commands remove the selected row(s) from your view or makes the row(s) appear on screen. The Unhide… command presents a dialog box for selecting the rows you want redisplayed.
	Column	Opens a submenu to control the size of column cells. The Width… option opens a dialog box in which you can define the width. The AutoFit Selection command adjusts the width to the size of the widest entry. The Hide and Unhide… commands perform as described in the Row command above. The Standard Width… command opens a dialog box to let you define the width of all the columns in the worksheet.
	Sheet	Presents a submenu to Rename… the worksheet as well as to Hide the current worksheet or Unhide… a worksheet.
Tools	Spelling…	Presents the Spelling tool, which you can use to search through the worksheet and have Excel check the spelling of words against a selected dictionary. You can choose to change spelling Excel finds inconsistent with the dictionary.
	Protection	Presents a submenu from which you can choose to protect the worksheet (Protect Worksheet) or the entire workbook (Protect Workbook) from changes.
	Options…	Presents the Options dialog box from which you can control a variety of Excel program options by making chooses from ten tabs.
Window	New Window	Creates a new window with a copy of the current worksheet.
	Arrange…	Presents the Arrange Windows dialog box from which you can choose to organize the windows of the various open workbooks for best reference and display.
	Hide	Hides the current window from view.
	Unhide	Presents the Unhide dialog box from which you can choose hidden windows to display once again.
	Split/Remove Split	Toggles between a Split view of the window which displays column header or row label information while you scroll through the worksheet or not showing the column header or row label data.
	Freeze Panes	Prevents the headers above or the labels to the left of the selected cell(s) from scrolling as you move through the worksheet.
	1, 2, 3, Command	Lists each available window by name. You can switch to another window by entering its number.
Help	Contents	Opens the Excel help file at its opening introductory page.
	Search for Help on…	Opens the Excel help file and its Search dialog box.
	Index	Opens the Excel help file to its index from which you can select from topics alphabetically.
	Quick Preview	Opens a guided tour of Excel to introduce you to Excel's use and features.
	Examples and Demos	Opens a list of examples and demonstration files you can choose to better understand Excel's use and features.
	Technical Support	Opens Excel help to topics that can connect you to technical support sources.
	About Microsoft Excel…	Presents a dialog box informing you about the current product and its installation.

Appendix II: Lotus 1-2-3 for Windows Menus and Commands

The following table lists the menus and their commands for Lotus 1-2-3 Release 5 for Windows that are different from Excel 4.0 for Windows (see Appendix I above). Not all the commands are included. Ones not included are those that have no relevance to this publication and which are not expected to be of immediate use to a new spreadsheet user. Be aware that different versions of programs will have different menu structures but the functionality will be available.

***Tip**: Lotus 1-2-3 offers many keyboard shortcut methods for issuing common commands. You can discover many of these by studying the menu and command entries next to which you can see the "hotkeys" for many commands. Other shortcut keyboard methods can be found in the Lotus 1-2-3 help file under the Keyboard Shortcuts topic.*

Menu	Command	Description
File	Doc Info...	See Appendix I File, Summary Info command.
	Protect...	See Appendix I Tools, Protection command.
Edit	Insert...	See Appendix I Insert Row, Column, Sheet commands.
	Find & Replace...	See Appendix I Edit, Search and Replace commands.
View	Freeze Titles...	See Appendix I Window, Freeze Panes command.
	Split...	See Appendix I Window, Split command.
	Set View Preferences...	Presents the Set View Preferences dialog box within which you can select options for how the program presents the current spreadsheet.
Style	Number Format..., Font & Attributes..., Alignment...,	See Appendix I Format, Cells command.
	Gallery...	Pick from an assortment of format templates for formatting the selected cell(s).
	Protection...	Presents the Protection dialog box for choosing to protect or not protect a cell (to protect the sheet, choose the File, Protect command).
	Hide...	See Appendix I Window, Hide and Unhide....
	Page Break...	See Appendix I insert, Page Break.
	Worksheet Defaults...	Presents a dialog box for setting cell font and attribute defaults for the whole spreadsheet.
	Fast Format	Copies the selected cell's style to paste onto and thus set for other cell(s).
Tools	Chart...	Presents the Chart dialog log box from which you can define a graph and chart.
	Spell Check...	See Appendix I Tools, Spelling.
	User Setup...	Presents the User Setup dialog box for setting various program options that affect how the program performs.
Range	Fill...	See Appendix I Edit, Fill.
	Transpose...	See Appendix I Edit, Paste Special....
	Name...	See Appendix I Insert, Name.
Window	Tile, Cascade, 1, 2, 3 Command	See Appendix I
Help	Contents, Search, etc.	See Appendix I